Introduction to the United States Air Force

B. CHANCE SALTZMAN, Capt, USAF
and
THOMAS R. SEARLE

Airpower Research Institute,
College of Aerospace Doctrine, Research and
Education, and Air University Press
Maxwell AFB, Alabama

2001

Library of Congress Cataloging-in-Publication Data

Saltzman, B. Chance, 1969–
    Introduction to the United States Air Force / B. Chance Saltzman, and Thomas R. Searle.
       p.   cm.
    ISBN 1-58566-092-2
    1. United States. Air Force. I. Searle, Thomas R., 1960– II. Title.

    UG633.S25 2001   2001046092
    358.4'00973—dc21

First Printing July 2001
Second Printing August 2001
Third Printing January 2003
Fourth Printing September 2004

Disclaimer

Air University Press
131 West Shumacher Avenue
Maxwell AFB AL 36112-6615
http://aupress.maxwell.af.mil

Foreword

The initial concept of the *Introduction to the United States Air Force* was to facilitate the process of learning how the US Air Force became what it is today: The most powerful military force in the history of the world. And as our Air Force continues to grow, so will this "primer." I wish to thank Tom Searle at CADRE for giving me the opportunity to help him update this introduction from an "end-user" point of view. This revised edition will take the student up to the twenty-first century by including some recent operations, aircraft, and significant personalities that were not included in the 1999 edition. Additionally, the 1999 edition included separate sections on significant operations and personalities but in this revised edition operations and personalities have been merged and organized chronologically to better match the needs of AS200. The original organization of the section on aerospace craft has been retained to show the unique developments of each of the different types of platforms.

For ROTC AS200 instructors: This book is intended to supplement *The Concise History of the United States Air Force* which is currently being used in the AS200 curriculum. Assigned reading from these books are highly encouraged to ensure the student has a firm grasp of the lessons prior to class.

Eric M. Moody, Capt, USAF
AFOATS Curriculum Development

Authors' Preface

To lead the US Air Force into the future, it is necessary to understand the past and present nature of the force. With this in mind, Air Force leaders have always sought to arm members of the force with a basic knowledge and understanding of Air Force culture and history. This volume is a contribution to that ongoing educational process, but as the title states, this is only an introduction. The information provided here merely scratches the surface of the fascinating stories of the people, equipment, and operations of the Air Force. Topics that are covered here in only a few short paragraphs have been, and will continue to be the subject of entire books. We hope this volume will be a starting point and a reference work to facilitate your continuing study of aerospace power.

The reader should keep in mind that all the people, operations, and aerospace craft included in this book have been important to the US Air Force, but they are not the only ones that have been important. The US Air Force has gained much from other nations, other US military services, and civilian organizations and these outside influences on the US Air Force are not included in this volume.

This *Introduction to the United States Air Force* is organized into two parts and five appendices. The first part is organized chronologically and groups significant operations and personalities together in several critical periods in the development of the US Air Force. The second part covers aerospace craft and is organized by type (fighters, bombers, missiles, etc.) in order to show the development of each type over time. Following Part II are appendices listing the senior leaders of the early air forces (before the creation of the US Air Force in 1947), the Air Force Chiefs of Staff, the Chief Master Sergeants of the Air Force, Fighter Aces, and Medal of Honor Winners.

While the biographical sketches and brief descriptions of operations are fairly self-explanatory, the information provided on aerospace craft requires some explanation. The dates given for aircraft and missiles are the initial delivery of production models, so prototypes will typically have been available earlier, and it may have been some years later before large operational units were fully trained and equipped with the new system. For satellites the date is the first successful launch of a satellite in the system. Aircraft types are continuously improved during their service lives so range, speed, and ceiling will vary widely between different models of the type. The statistics listed in this book are typical for the aircraft but on a specific day, actual performance of any single aircraft may have been different. Aircraft range is particularly difficult to specify because it can be changed by using external fuel tanks of various sizes and aerial refueling can make range virtually infinite. Satellite systems pose problems of their own due partly to security classification. Additionally, satellite systems are listed by function rather than model and over a period of decades, continuous improvement in the satellites performing that function will result in changes in the size, shape, and weight of the satellites. Thus the pictures of satellites and the dimensions given for them represent typical recent satellites performing that function, but may be quite different from past and future systems. For all these reasons, the statistics provided in Part I must be considered simplified and incomplete synopses of much more complicated stories.

No work of this type could have been written without the invaluable assistance of many people and institutions. We would like to take this opportunity to thank a few of them. Dr. Daniel Mortensen and Col (Ret.) Phillip S. Meilinger are two outstanding airpower scholars who gave us enormous assistance and encouragement. At Air University Press, John Jordan, Peggy Smith, Daniel Armstrong, and Mary Ferguson provided the editing, layout, and photograph selection without which this volume would have been a mess. Any faults that remain are, of course, entirely our own.

B. Chance Saltzman and
Thomas R. Searle

Contents

Page

FOREWORD . . . . *iii*

AUTHORS' PREFACE . . . . *v*

Part I

WORLD WAR I AND THE INTERWAR YEARS: 1916–40 3
    The Mexican Punitive Expedition (1916–17) 5
    St. Mihiel Offensive of World War I (September 1918) 6
    The Sinking of the *Ostfriesland* (1921) 8
    The Pan American Goodwill Flight (1927) 9
    The *Question Mark* Flight (1929) 10
    The Airmail Operation (1934) 11
    "Hap" Arnold's B-10 Flights to Alaska (1934) 12
    Maj Gen Benjamin D. "Benny" Foulois 13
    Eugene Jacques Bullard 14
    Brig Gen William "Billy" Mitchell 15
    Capt Edward V. "Eddie" Rickenbacker 16
    Lt Frank Luke Jr. 17
    Maj Gen Mason M. Patrick 18
    Lt Gen Frank M. "Andy" Andrews 19

WORLD WAR II: 1941–45 21
    The Doolittle Raid on Tokyo (18 April 1942) 23
    Kasserine Pass (February 1943) 25
    Battle of the Bismarck Sea (1943) 26
    Ploesti Raid (August 1943) 27
    The Regensburg/Schweinfurt Raids (1943) 28
    "The Big Week" (February 1944) 29
    Operation Overlord (1944) 30
    Bombing Tokyo (9–10 March 1945) 31
    The Hiroshima and Nagasaki Bombings (August 1945) 32
    General of the Army Henry H. "Hap" Arnold 33
    Lt Gen Claire L. Chennault 34
    Lt Gen Ira C. Eaker 35
    Gen Carl A. "Tooey" Spaatz 36
    Col Francis S. "Gabby" Gabreski 37
    Capt Colin P. Kelly Jr. 37
    Gen Laurence S. Kuter 38

Page

Maj Richard I. "Dick" Bong 39
Lt Gen James H. "Jimmy" Doolittle 40
Gen George C. Kenney 41
Lt Gen Ennis C. Whitehead 42
Lt Gen Benjamin O. Davis Jr. 43
Jacqueline "Jackie" Cochran 44
Gen Leon W. Johnson 45
Maj Thomas B. "Mickey" McGuire Jr. 46
Brig Gen Paul W. Tibbets Jr. 47

INDEPENDENT AIR FORCE, COLD WAR, AND KOREA: 1946–64 49
X-1 Supersonic Flight (14 October 1947) 51
Berlin Airlift (1948–49) 52
Pusan Perimeter Defense (1950) 53
Corona (1960) 54
Strategic Air Command (1946–92) 55
The Cuban Missile Crisis (1962) 56
Brig Gen Charles E. "Chuck" Yeager 57
Gen Hoyt S. Vandenberg 58
Col Dean Hess 59
Col James Jabara 60
Capt Joseph C. "Joe" McConnell Jr. 60
Maj Gen Frederick C. "Boots" Blesse 61
Brig Gen Robinson "Robbie" Risner 62
Gen Nathan F. Twining 63
Lt Gen William H. Tunner 64
Gen Curtis E. LeMay 65
Gen Bernard A. "Bennie" Schriever 66
Lt Col Edwin "Buzz" Aldrin 67
Lt Col Virgil I. "Gus" Grissom 67
Maj Donald K. "Deke" Slayton 67
Chief Master Sergeant of the Air Force Paul W. Airey 68

THE VIETNAM WAR: 1965–75 69
Operation Rolling Thunder (1965–68) 71
Operation Arc Light (1965–67) 72
Operation Bolo (January 1967) 73
Operation Linebacker I (1972) 74
Operation Linebacker II (1972) 75
A1C William H. Pitzenbarger 76
Lt Col Merlyn H. Dethlefsen 77
Capt Lance P. Sijan 78
Gen John D. Ryan 79
Gen Daniel "Chappie" James Jr. 80
A1C John L. Levitow 81

Gen George S. Brown 82
Capt Richard S. "Steve" Ritchie 83
Gen David C. Jones. 84
Brig Gen Robin Olds 85
Brig Gen Harry C. "Heinie" Aderholt 86

REBUILDING THE AIR FORCE AND THE GULF WAR: 1976–91

REBUILDING THE AIR FORCE AND THE GULF WAR: 1976–91 87
El Dorado Canyon (14–15 April 1986) 89
Operation Just Cause (December 1989) 90
Strategic Attack on Iraq (January 1991) 91
The Battle of Khafji (1991) 92
Col John W. Warden, III 93
Gen Charles A. Horner 94
Gen Merrill A. McPeak 95

POST-COLD WAR: 1992–2000

POST-COLD WAR: 1992–2000 97
Policing Postwar Iraq (1991–?) 99
Operations in Somalia (1992–94) 101
Operations Provide Promise and Deny Flight: Bosnia (1992–96) 102
Operation Uphold Democracy: Haiti (1994) 104
The Kosovo Crisis (1999) 105
Gen Ronald R. Fogleman 107
Lt Gen Michael C. Short 108

Part II

EARLY AIRCRAFT

EARLY AIRCRAFT 111
Wright 1909 Military Flyer and Curtiss JN-4D "Jenny". 111
DeHavilland DH-4 and SPAD XIII 112

PURSUIT/FIGHTER AIRCRAFT

PURSUIT/FIGHTER AIRCRAFT 113
Curtiss P-6E and Boeing P-26 113
Bell P-39 and Lockheed P-38 114
Curtiss P-40 and Republic P-47 115
North American P-51 and Bell P-59 116
Northrop P-61 and Lockheed F-80 117
Republic F-84 and North American F-86 118
North American F-100 and McDonnell F-101 119
Convair F-102, Lockheed F-104, and Convair F-106 120
Republic F-105 and McDonnell Douglas F-4 121
General Dynamics F-111 and McDonnell Douglas F-15 122
Lockheed Martin F-16 and Lockheed F-117 123
Boeing and Lockheed Martin F-22 124

BOMBERS 125
   Martin GMB/MB-1, Boeing B-9, and Martin B-10 125
   Boeing B-17 and Consolidated B-24 126
   North American B-25 and Martin B-26 127
   Boeing B-29 and Convair B-36 128
   Boeing B-47 and Convair B-58 129
   Boeing B-52 and Rockwell B-1 130
   Northrop Grumman B-2 131

ATTACK AIRCRAFT 133
   Northrop A-17 and Douglas A-20 133
   Douglas A-26 and Douglas A-1 134
   Vought A-7 and Fairchild A-10 135

TANKERS 137
   Boeing KC-97 and Boeing KB-50 137
   Boeing KC-135 and McDonnell Douglas KC-10 138

TRANSPORT AND SPECIAL ELECTRONIC AIRCRAFT 139
   Curtiss-Wright C-30 and Douglas C-47 139
   Curtiss C-46 and Douglas C-54 140
   Fairchild C-82/119 and Douglas C-124 141
   Lockheed C-69/121 and Fairchild C-123 142
   Lockheed C-130 and Lockheed C-141 143
   Lockheed C-5 and Boeing E-3 144
   Northrop Grumman E-8 and Boeing C-17 145

TRAINING AIRCRAFT 147
   Vultee BT-13 and North American AT-6 147
   Lockheed T-33 and North American T-28 148
   Beech T-34 and Cessna T-37 149
   Northrop T-38 and Raytheon T-1 150

EXPERIMENTAL AND STRATEGIC RECONNAISSANCE AIRCRAFT 151
   Bell X-1 and Lockheed U-2 151
   North American X-15 and North American XB-70 152
   Lockheed SR-71 153

STRATEGIC MISSILES AND SPACE LAUNCH VEHICLES 155
   Douglas PGM-17 Thor and General Dynamics HGM-19 Atlas 155
   Martin LGM-25 Titan and Boeing LGM-30 Minuteman 156
   Boeing LGM-118 Peacekeeper and McDonnell Douglas Delta II 157
   Boeing AGM-86 and Boeing AGM-129 Advanced Cruise Missile 158

SATELLITES 159

    Lockheed Corona 159

    Lockheed Defense Meteorological Satellite Program

      and Defense Satellite Communications System 160

    TRW Defense Support Program and Rockwell Navstar

      Global Positioning System 161

    Milstar Satellite Communications System 162

Appendix

A Senior Leadership of the Early Air Forces 165

B Air Force Chiefs of Staff 167

C Chief Master Sergeants of the Air Force 169

D Fighter Aces 171

E USAF Medal of Honor Winners 175

Part I

Major Operations and Personnel of the United States Air Force

World War I and the
Interwar Years

1916–40

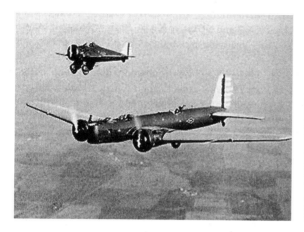

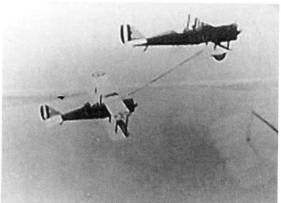

The Mexican Punitive Expedition (1916–17)

1st Aero Squadron JN-3 in Mexico.

Woefully inadequate equipment and poorly trained personnel limited the airmen's contribution to delivering dispatches and mail.

In 1911 the Mexican government was overthrown and the country descended into civil war. An incident in Vera Cruz in 1914 led to US Navy intervention. Cross border raids by Mexican revolutionaries (which killed nine soldiers and six civilians) and assaults on American citizens and property in Mexico further heightened tensions. In January 1916 forces loyal to the bandit/revolutionary Pancho Villa, deliberately killed 18 Americans in Mexico, and in March, they crossed the border and attacked the border town of Columbus, New Mexico, killing eight soldiers and eight civilians. President Woodrow Wilson ordered American forces under the command of Gen John Pershing to pursue and attack Villa's forces in Mexico. On 15 March 1916 American troops crossed the international border with grudging permission from the Mexican government. Pershing's force included the newly-formed 1st Aero Squadron with 8 aircraft, 10 pilots, 84 enlisted men, 10 trucks, one automobile, and six motorcycles.

The airmen should have been a great help to the cavalry units operating against a very mobile foe in difficult terrain. They should have been ideal for reconnaissance and even in small numbers, their attacks should have had a major impact on the campaign. Unfortunately, poor equipment and woefully inadequate training severely limited the role of the Aero Squadron. The squadron was equipped with the Curtiss JN-3 which was adequate as a trainer but could not cross the 10,000-foot mountains in that part of Mexico, could only carry a payload of 265 pounds, had no instruments, and was unarmed. The pilots were no better prepared than their aircraft and the high accident and incident rate was not a surprise. Maintenance was a further problem because the wood and canvas suffered from the desert climate (especially the wooden props). After a month of operations only two of the original eight aircraft were still operational, and they were condemned.

The squadron received Curtiss NBs and other new aircraft and the pilots got a lot of on-the-job training but the squadron's main contribution to the campaign was carrying dispatches and mail. Their most significant single accomplishment was to find a lost cavalry column. The weaknesses of the United States air arm rapidly became clear to everyone, as did many of the requirements for conducting a sustained air campaign. The steps taken to remedy these problems bore fruit less than two years later when the 1st Aero Squadron and the rest of the American Air Service demonstrated their combat capabilities in World War I.

St. Mihiel Offensive of World War I (September 1918)

Brig Gen William "Billy" Mitchell

O n 12 September 1918, American forces under Gen John J. Pershing, began the offensive to reduce the St. Mihiel salient. The bulge was a 24-mile wide, 14-mile deep salient formed by the German advance south of the fortress city of Verdun earlier in the war. For the offensive, General Pershing commanded the Allied First Army in an excellent example of a joint and combined operation. In addition to US Army air and ground units, the US ground contingent of Pershing's force included a large number of US Marines. The Allies also contributed substantial forces. One of Pershing's four ground corps was made up of French Colonial troops and air units were provided by France, Italy, Portugal, and Great Britain. The American units were under what we would now call Operational Control of General Pershing, but the British air units came from the "Independent Force." They coordinated and cooperated with the St. Mihiel offensive, but in an early example of the complex command relationships in combined operations, they were not under Pershing's control.

St. Mihiel represented the first large scale massing of airpower under central command. With 1,500 aircraft, General Mitchell seized the initiative, gained air superiority, attacked ground forces, and interdicted supplies, helping Allied ground forces to achieve their objectives.

General Pershing appointed Col (later Brig Gen) William "Billy" Mitchell as his air commander for the battle. Mitchell's role was similar to that of a modern joint force air component commander (JFACC) in that he was both commander of air forces and the chief advisor to the overall commander (in this case General Pershing) on the employment of air assets. Mitchell's forces, consisting of airplanes and observation balloons, were the largest massing of airpower the world had yet seen. About 600 of these planes were flown by US personnel.

St. Mihiel represented the first large scale massing of airpower under central command. With 1,500 aircraft, General Mitchell seized the initiative, gained air superiority, attacked ground forces, and interdicted supplies, helping Allied ground forces to achieve their objectives.

In 1918, the Allies organized their aircraft into four types: observation (which we now call reconnaissance and surveillance); pursuit (which we now call fighters); daylight bombers; and night bombers. With 1918 technology, the artillery of the ground armies could deliver vastly more firepower than the small number of primitive bombers available to Colonel Mitchell, so a large portion of the observation force (and all of the observation balloons) provided targeting data to the ground artillery. To facilitate the passing of data between the artillery and aerial reconnaissance units, many of these units were assigned to support the ground commanders and their artillery units. General Mitchell kept all of the pursuit and bombardment planes under his direct control and retained many of the observation units as well.

Mitchell believed that the first task of airpower was achieving air superiority and he was concerned that ground commanders would try to use his aircraft defensively to patrol over friendly lines and protect ground units and aerial observers from German planes. Instead, Mitchell was determined to use his pursuit planes in offensive counterair operations to attack the Germans deep in their own territory and drive them from the air while the bombers attacked targets deep in the German rear. Mitchell's approach was extremely successful and helped the Allies gain and maintain control of the air throughout the offensive. In a foreshadowing of "Big Week" and other air campaigns of World War II, German pursuit aircraft came up to attack Allied bombers and observation planes and were destroyed by Allied pursuit aircraft that were hunting for enemy aircraft rather than merely defending friendly ones.

While observation operations focused on areas near the front lines, the 91st Aero Squadron conducted some deeper operations. This unit was particularly effective because it was one of the most experienced units in the American Air Service and had been operating in the St. Mihiel district since the spring. The bombers focused on the counterland campaign with close air support missions and interdiction missions against road and rail junctions in the German rear. The bombers operated in larger formations to defend themselves and free the Allied pursuit aircraft to hunt German planes and balloons.

During the four days of the offensive, American airmen made 3,300 flights over battle lines, logged 4,000 hours, made 1,000 individual bomb attacks dropping 75 tons of high explosive. They also destroyed 12 enemy bal-

Capt Eddie Rickenbacker with his Spad XIII, briefly took the lead as the "Ace of Aces" on 15 September with his 8th victory. Later that day, Lt Elliot Springs scored his 9th victory to reclaim the lead.

loons and more than 60 enemy planes not to mention the enormous damage done by strafing German units trying to escape along the roads out of the salient. In this brief battle the First Army cleared the St. Mihiel salient and captured 15,000 prisoners and more than 250 cannons. Massed airpower contributed substantially to this achievement. In General Pershing's words, "The organization and control of the tremendous concentration of Air Forces has enabled the First Army to carry out its dangerous and important mission."

In addition to famous US aces like Eddie Rickenbacker and Frank Luke, US airmen in the St. Mihiel offensive included George C. Kenney, Joseph T. McNarney, and Carl A. Spaatz. Kenney, McNarney, and Spaatz all went on to become four-star generals in the US Air Force and Spaatz became the first Chief of Staff of the US Air Force. Rickenbacker, Luke, and Spaatz all earned aerial victories during the St. Mihiel air battles and Luke scored 8 of his 18 victories there. More importantly, many of the future leaders of US airpower gained invaluable experience in large-scale air operations at St. Mihiel. This experience paid immediate dividends in World War I and was invaluable to the future development of US airpower.

7

The Sinking of the *Ostfriesland* (1921)

An MB-2 hits its target, the obsolete battleship USS *Alabama* during tests.

A Martin MB-2, an early version of the bombers used in the tests.

Following World War I, many Air Service flyers supported the idea of an air arm, independent of the Army and Navy, and felt that aircraft could protect US shores more efficiently and effectively than the battleships of the US Navy that had traditionally done that job. The most prominent advocate of this philosophy was Billy Mitchell.

When Mitchell suggested that US airpower could defend the nation's coasts from attacks by enemy warships better than US sea power, a controversy developed as to whether an airplane could even sink a battleship. No battleship had ever been sunk by an airplane and the Secretary of the Navy, Josephus Daniels, expressed the Navy's contempt for Mitchell when he offered to stand bareheaded on the bridge of any ship the Army attempted to bomb. He approved tests off the Virginia capes to be conducted in June and July of 1921. During the tests, Mitchell's bombers, MB-2s of the 1st Provisional Brigade flying from Langley Field, Virginia, sank three captured German naval vessels including the *Ostfriesland*, a fairly modern battleship built by the Germans

only 10 years before. In September 1921, the obsolete United States Ship (USS) *Alabama* was sunk. Two years later, additional tests were conducted off Cape Hatteras and two more obsolete US battleships were sent to the bottom. The Navy argued that the tests had been conducted under artificial conditions. They contended that damage-control parties could have stopped the flooding that sank the *Ostfriesland* and claimed that anti-aircraft fire could have eliminated several of the bombers before they could strike. Nevertheless, bombs from aircraft had sunk battleships and a realization that the Navy would require air cover in future wars began to spread. This awareness was a boon to Navy airmen who began to receive much more attention as a result of these tests.

For Mitchell, the sinkings in 1921, and in subsequent tests, proved conclusively that the day of the battleship had ended and that an independent air force could defend the coasts of the United States and its possessions more effectively and less expensively than any combination of the Army's coast artillery and the Navy's warships. The Navy agreed that battleships were vulnerable when they came within range of bombers, but they still firmly believed that the dreadnought reigned supreme on the high seas beyond reach of land-based aircraft.

The success of the bombing trials encouraged the supporters of a separate air arm to press even harder for their objectives but the Army General Staff remained firm in its belief that airpower, acting independently, could not win a war. For its part, the US Navy became even more adamant that it, not an independent air force, should control naval aviation. The debate over proper and effective organization and use of airpower began before the United States entered World War I and continues to this day.

The Pan American Goodwill Flight (1927)

Lt Gen Ira C. Eaker

In the mid-1920s, very little flying had been done in Latin America and anti-US feelings were strong there in spite of the long-standing US strategic interest in the region. Latin Americas were also buying European airplanes rather than American ones. In the hopes of garnering good publicity for the US Army Air Corps, opening up Lating America to military and commerical aviation, and interesting Latin Americans in US-built airplanes, the War Department authorized the Pan American Goodwill Flight. The flight started in San Antonio, Texas, and proceeded south through Mexico and along the Pacific coast of Latin America before crossing the continent in southern Chile and returning to the US along the Atlantic coast of South America and across the West Indies to Washington, D.C.

Maj Gen James Fechet, deputy chief of the Air Corps nominated 20 officers to Maj Gen Mason M. Patrick, chief of the Air Corps. Patrick appointed Maj Herbert A. Dargue as commander and nine other pilots, including Capt Ira C. Eaker to fly the mission. Eaker's copilot was Lt Muir S. Fairchild. The Pan American flights were highly publicized and captured world attention. Eaker and Fairchild won particular glory because their OA-1, *San Francisco*, was the only one that completed the entire 23,000-mile journey, making every scheduled stop.

The flight received front-page coverage almost every day for five months in major US newspapers and many foreign ones. The purposes of the flight included furthering relations with Latin American countries, encouraging commercial aviation, providing valuable training for Air Corps personnel, and giving an extensive test to the amphibian airplane. The OA-1 amphibian was designed for observation work. The engine was a water-cooled Liberty engine producing 400 horsepower, mounted upside down so that the three-bladed aluminum propeller could clear the hull's upturned beak. Each aircraft was named after an American city to further promote relations.

The flights did have their share of trouble. The *New York* crash-landed in Guatemala and the *San Antonio* required an engine change in Columbia. Tragedy struck when two aircraft collided over Buenos Aires. Major Dargue and Lt Ennis C. Whitehead escaped by parachute but Captain Woolsey and Lieutenant Benton went down with the *Detroit* and were killed instantly.

Despite these setbacks, the Pan American flights were a success. Upon arrival at Bolling Field, Washington, D.C., the fliers were met by President Calvin Coolidge who presented each of them with the first Distinguished Flying Crosses, a medal authorized by Congress a few months before. The Americans had aroused the aviation interest of Latin Americans, many of whom had never seen an airplane. The expedition maps became the basis for civil aviation routes in Latin American.

The OA-1 amphibious aircraft, the *San Francisco,* flown 23,000 miles by Ira Eaker and Muir Fairchild.

9

The *Question Mark* Flight (1929)

The *Question Mark* Team, 1929.

The record-setting Question Mark *flight demonstrated effective air-to-air refueling and the range possibilities of aircraft. Bombers would be able to seek out distant targets and the oceans would be eliminated as barriers.*

In March 1928, Captain Eaker and Lt Elwood R. "Pete" Quesada were involved in the rescue of the three-man crew that accomplished the first east-west flight across the Atlantic. The experience left Eaker and Quesada with an idea—an effective way of refueling in midair. Captain Eaker was just the man to turn this project into reality. It took a mixture of imagination, political savvy, opportunism, and zeal to succeed. At Bolling Field, he spotted two suitable aircraft, a trimotored Fokker and a Douglas. Eaker then assembled his team. He enlisted the help of Maj Carl A. "Tooey" Spaatz and suggested that he be the commander of the project. Eaker would act as chief pilot with Quesada as copilot. The C-1 crew included Lt Harry Halverson and Sgt Roy Hooe.

On 1 January 1929, the *Question Mark,* so named because the crew did not know how long it would be airborne, took off from San Diego, California. It flew back and forth over the 70 miles between Los Angeles, California, and San Diego for six days. Major Spaatz rode in the tail section of the airplane and was responsible for connecting the refueling hose to the tank. The refuelings were dangerous with the C-1 hovering only 20 feet above the Fokker at all hours of the day and night and through turbulence. When the hose was lowered, Spaatz would ground it to a copper plate fixed to the Fokker to prevent a spark and then fit it in the tank's funneled open-ing. The gas would flow at 75 gallons per minute. Oil in five-gallon containers, food, and supplies were lowered from the C-1 by rope and hauled into the Fokker by Spaatz.

On 7 January the Fokker's left engine quit and although the plane could remain aloft on one engine it could not maintain the five thousand feet deemed safe. After 150 hours, 40 minutes, and 15 seconds, the *Question Mark* touched down. It had flown about 11,000 miles and set an endurance record that would last for many years. The contributions of this flight were numerous. Besides the obvious air-to-air refueling possibilities, the flight also showed that a plane powered by 1929 engines could fly 11,000 miles with five men aboard. With new planes already on the drawing board, regular nonstop transcontinental flights would soon be practical. The flight also foreshadowed American bombers seeking distant targets and likewise that the United States would soon be vulnerable to aerial attack. Finally, the ocean was eliminated as a barrier and the Navy's wartime role was forever altered.

The *Question Mark* refueling in flight.

The Airmail Operation (1934)

A Curtiss B-2 Condor over Airmail Route 4 (the Salt Lake City—Los Angeles mail).

In February 1934, President Franklin D. Roosevelt canceled the airmail contracts between the Post Office and the commercial airlines due to evidence of fraud in the conduct of the business. Before issuing a formal cancellation directive, Roosevelt had postal officials meet with the Chief of the Air Corps, Maj Gen Benjamin D. "Benny" Foulois, to determine if the Air Corps could temporarily haul the mail. Foulois agreed to the proposal because the publicity of the successful operation would be helpful in securing badly need funds for the Air Corps.

On 9 February 1934, Foulois committed the Air Corps to the task of delivery the nation's airmail. The endeavor proved to be too much for the Air Corps. The airmail flights took place at night and regardless of weather conditions; however, the average Army pilot lacked experience in this kind of flying. The Army's open cockpit planes were not normally equipped with blind flying instruments or even radios. In addition to the equipment and training shortcomings, the staff was faced with a huge administration challenge. Hangars had to be acquired, office space along the routes secured, refueling arrangements made, and scheduling and maintenance logistics all had to be sorted out. The task was truly daunting. The mail was carried in every available type of aircraft. Twenty-five pounds could be carried in the P-12, while two thousand pounds could be hauled in the Martin B-10 and Boeing B-9.

Maj Gen Oscar M. Westover, deputy chief of the Air Corps, took over the airmail operation and established geographic zones. The western zone was commanded by "Hap" Arnold, the central by Horace Hickam, and Byron Q. Jones commanded the eastern zone.

The Army Air Corps began flying the mail on 19 February 1934. Unfortunately, the United States experienced some of the worst late winter weather in its history. Throughout February, rain, snow, dense fog, and icy gales covered the nation. Aircraft on ramps couldn't be started, carburetor icing caused engine failure, and pilots became disoriented in fog and clouds. Valor and eagerness could not overcome lack of experience and poor weather. Six Army pilots were killed in the operation's first week. A frustrated President Roosevelt took steps to renegotiate with the commercial airlines.

By March, the weather was improving and a quick course in instrument flying combined to improve the Air Corps' record. However, the overall effort was dismal. In the 12 months that the Air Corps delivered mail, there were 66 crashes, 12 deaths, and only a 66 percent sortie completion rate. During this period, the number of deaths rose 15 percent from the average.

Despite its poor showing, the Air Corps learned immensely from its airmail operation. The Air Corps learned there was a need for an extensive instrument training program, an automated landing system, and reliable radio communications. The airmail fiasco also led to the convening of the Drum Board, which recommended the creation of General Headquarters (GHQ) Air Force. Followed by the Baker Board that supported the formation of a centrally controlled aerial strike force. The findings of these boards established aviation policy that endured through World War II, acknowledging that military aviation was valuable in both offensive and defensive operations.

"Hap" Arnold's B-10 Flights to Alaska (1934)

General of the Army "Hap" Arnold

The B-10 Alaska flights proved the strategic application of the Army's bombers.

In the summer of 1934, Lt Col Henry H. "Hap" Arnold was ordered to Wright Field in Dayton, Ohio, to organize a flight of bombers to Alaska. The planes to be used were the Martin B-10s. The B-10 was a low-wing monoplane, the Air Corps' first all-metal bomber. It was also the first 200 miles per hour (MPH) bomber with a range of 900 miles. The flight path would take the bombers from Dayton, Ohio, to Washington, D.C. From Washington, D.C., the bombers would travel to Minneapolis, Minnesota, into Canada across Alberta to Prince George then to Fairbanks, Alaska. The mission would include aerial photography of Alaska to map airways in and out of Alaska and to various ports of entry to Alaska from the Soviet Union. The return flight would send the bombers south from Juneau, Alaska, to Seattle, Washington, then on to Washington, D.C. The flight would include 10 bombers with 14 officers and six enlisted men. After a month of preparation, the B-10s took off for Washington on 18 July 1934. The mission was without precedent but was not unusually tough. The B-10s averaged 820 miles a day except from Juneau to Seattle where there was no landing field for 990 miles. The mission covered 18,000 miles over very rough terrain without a major incident. In Cooks Bay (Anchorage), Alaska, one of the B-10s sank into 40 feet of water. It was hauled out and repaired and continued the mission. Arnold's crew chief, TSgt Henry Puzenski, was largely responsible for this miracle and remained Arnold's crew chief until he retired 13 years later. The B-10s returned to Washington, D.C., landing on 22 August 1934. The return trip angered Army Chief of Staff Gen Douglas MacArthur because Arnold had flown his B-10s off course over the ocean to demonstrate their coastal range. This grandstanding angered the Navy and MacArthur had to deal with the delicate situation centering on which service was responsible for coastal defense. Arnold's detour ensured that no medals were given out for the Alaska flight as long as MacArthur was Chief of Staff. Despite the politics, the trip had proven the extensive range and strategic application of the Army's bombers.

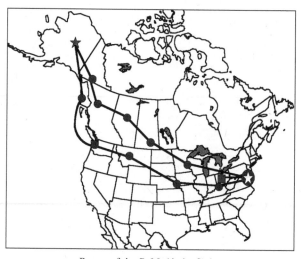

Route of the B-10 Alaska flights.

Maj Gen Benjamin D. "Benny" Foulois

- Taught himself to fly largely through correspondence with the Wright brothers in 1910.
- During World War I, Gen "Benny" Foulois was responsible for all Air Service support functions.
- After serving as air attache in Germany and commanding Mitchel Field in New York, General Foulois was named Chief of the Air Corps in 1931.
- As chief of the Air Corps, Foulois was largely responsible for the "Air Mail Fiasco" of 1934. Delivering the country's mail proved to be a task the young air force was not equipped to undertake.

Benjamin D. Foulois taught himself to fly largely through correspondence with the Wright brothers in 1910. While many of his contemporaries died in plane crashes or quit flying, he continued as an operational pilot until World War I.

Foulois came from a humble background and was physically unimpressive; worse, he lacked the charisma of his contemporary and chief rival within the air arm, William "Billy" Mitchell. The animosity between Mitchell and Foulois began in 1916 when he and the 1st Aero Squadron were sent to Mexico with John Pershing in pursuit of the bandit Pancho Villa. At the same time, the chief of the Signal Corps was forced from office due to financial improprieties, and Mitchell, who had not yet even flown an airplane, was temporarily placed in charge. The poor performance of the aviation unit in Mexico resulted in mutual finger pointing between Mitchell and Foulois, and the rift never healed. Pershing clearly had respect for both of them, but thought neither had the experience or maturity to run the Air Service; hence, he appointed Mason M. Patrick to lead the air arm and control its two main recalcitrants. Nonetheless, as chief of the Air Corps after 1931, Foulois's steady perseverance in working to shift War Department views regarding the importance of airpower gradually paid off, resulting in the increased autonomy of the General Headquarters (GHQ) Air Force, formed in 1935. On the other hand, Foulois was not popular among his Army brethren and the disastrous performance of the Air Corps in the "Air Mail Fiasco" of 1934 earned him the ill will of President Roosevelt. Looking for a scapegoat, Congress, also embarrassed by the miserable Air Corps performance, held hearings into the issue of aircraft procurement. Foulois was reprimanded for "misleading" Congress and violating the spirit of procurement laws. The Air Corps chief's relations with the Navy were also stormy during this period. But in truth—given the budget crunch during the bottom of the Depression, their inherently conflicting views regarding the role of airpower in war, and the poisoned atmosphere created by Mitchell—such difficulties were inevitable. Without friends in or out of the Army and with his usefulness clearly limited, Foulois retired in December 1935, a bitter and lonely man. But he lived to see the US Air Force become the most powerful military force in the world and he was a regular speaker at Air Force bases in the 1960s.

Eugene Jacques Bullard

- First African-American military pilot.
- Served in the French Army Aviation Service in World War I.
- Remained in Paris after the war and spied on Germany in World War II.

Of more than two hundred Americans who flew for France during World War I, one of particular uniqueness was Eugene Bullard, the only black pilot of World War I. A fellow flier describes Bullard in France, "His jolly face shone with a grin of greeting and justifiable vanity. He wore a pair of tan aviator's boots which gleamed with a mirror-like luster, and above his breeches smote the eye with a dash of vivid scarlet. His black tunic, excellently cut and set off by a fine figure, was decorated with a pilot's badge, a Croix de Guerre, the fourragere of the Foreign Legion, and a pair of enormous wings, which left no possible doubt, even at a distance of 50 feet, as to which arm of the Service he adorned. The eleces-pilotes gasped, the eyes of the neophytes stood out from their heads, and one repressed a strong instinct to stand at attention."

There was scarcely an American at Atord who did not know and like Bullard. He was a brave, loyal, and thoroughly likable man but the US Air Service did not accept black officers. Therefore, he fought as a member of the French Army Aviation Service and not as a member of the US Air Service.

Following World War I, Bullard remained in France until the German occupation of Paris in 1940, at which time he had to flee the country because of his previous spying activities against the Nazis. He returned to the United States and lived in New York City, New York, until his death in 1961. Thus passed from the scene the first black pilot in the history of military aviation.

Bullard flew the Spad XVI, a French-designed and -built fast two-seat reconnaissance-type aircraft developed from the Spad VII. At the time of its entry into service (1916), it was one of the fastest and most maneuverable two-seaters in combat.

14

Brig Gen William "Billy" Mitchell

- Mitchell illustrated the power of strategic bombing by sinking several battleships during tests in 1921 and 1923.
- In September 1918, Gen Mitchell planned and led nearly fifteen hundred allied aircraft in the air phase of the St. Mihiel offensive.
- As the top American combat airman of World War I, Gen Mitchell was awarded the Distinguished Service Cross, the Distinguished Service Medal, and several foreign decorations after his 18 months in France.
- In 1925 Mitchell was court-martialed and found guilty of insubordination. He resigned in 1926 and continued to promote airpower in his civilian life.

Billy Mitchell is the most famous and controversial figure in American airpower history. The son of a wealthy Wisconsin senator, he enlisted as a private during the Spanish–American War. Quickly gaining a commission due to the intervention of his father, he joined the Signal Corps. He was an outstanding junior officer, displaying a rare degree of initiative, courage, and leadership. After challenging tours in the Philippines and Alaska, Mitchell was assigned to the General Staff—at the time its youngest member. He slowly became excited about aviation, which was then assigned to the Signal Corps, and its possibilities, and in 1916 at age 38, he took private flying lessons.

Arriving in France in April 1917, only a few days after the United States had entered the war, Lieutenant Colonel Mitchell rapidly earned a reputation as a daring, flamboyant, and tireless leader. He eventually was elevated to the rank of brigadier general and commanded all American combat units in France. In September 1918 he planned and led nearly fifteen hundred Allied aircraft in the air phase of the St. Mihiel offensive. Recognized as the top American combat airman of the war (he was awarded the Distinguished Service Cross, the Distinguished Service Medal, and several foreign decorations),

Mitchell managed to alienate most of his superiors, both flying and nonflying, while in France.

Returning to the United States in early 1919, Mitchell was appointed the deputy chief of the Air Service, retaining his one-star rank. His relations with superiors continued to sour as he began to attack both the War and Navy Departments for being insufficiently farsighted regarding airpower. His fight with the Navy climaxed with the dramatic bombing tests of 1921 and 1923 that sank several battleships, proving, at least to Mitchell, that surface fleets were obsolete. Within the Army he also experienced difficulties, notably with his superiors, Charles Menoher and later Mason M. Patrick, and in early 1925 he reverted to his permanent rank of colonel and was transferred to Texas. Not content to remain quiet, when the Navy dirigible *Shenandoah* crashed in a storm and killed 14 of the crew, Mitchell issued his famous statement accusing senior leaders in the Army and Navy of incompetence and "almost treasonable administration of the national defense." He was court-martialed, found guilty of insubordination, and suspended from active duty for five years without pay. Mitchell elected to resign instead as of 1 February 1926, and spent the next decade continuing to write and preach the gospel of airpower to all who would listen.

15

Capt Edward V. "Eddie" Rickenbacker

- He had 26 confirmed aerial victories while engaging in 134 dog-fights during World War I.
- One of his eight Distinguished Service Crosses was upgraded to a Medal of Honor in 1930.
- He was the last witness for the defense in the court-martial of Gen Billy Mitchell in 1925.
- The American "Ace of Aces," went on to become president of Eastern Airlines before his death in 1973.

Capt Eddie Rickenbacker was originally turned down for enlistment for lack of education but was persistent, and on 25 May 1917 in New York City, he joined the Signal Enlisted Reserve Corps with an assignment to the Aviation Section. Three days later he was on his way to Paris, France, for assignment to Aviation Headquarters American Expeditionary Forces (AEF) as a sergeant first class. Partly because Rickenbacker had been a successful race car driver before the war, he was initially assigned as Gen John J. Pershing's staff driver. At his insistence he was permitted to join a fighter unit, being assigned as a student at the Aviation Training School at Tours, France. He completed the course 10 October 1917, and was commissioned a first lieutenant. He then became engineering officer to HQ Detachment, 3d Aviation Instruction Center at Issoudun, serving under Maj Carl A. Spaatz, whose own fame lay ahead in World War II. For many months Spaatz refused to allow him to become a combat pilot because he needed him on the staff. However, in March 1918, Captain Rickenbacker prevailed in his repeated requests and he was assigned to the 1st Pursuit Group's 94th Aero Squadron, the famed "Hat-in-the-Ring" Squadron, as a pilot under Maj Raoul Lufbery. Rickenbacker was in action the next month, flying his Nieuport fighter over the lines against the enemy on 25 April,

and shooting down a German flying a Pfalz, without a single bullet hitting his own machine.

By 1 June 1918, Rickenbacker had become an ace, with five enemy kills to his credit. A daring, fear-less, talented but never reckless flyer, Rickenbacker as commanding officer, would never assign his men to a target he wouldn't lead. He continued as Hat-in-the-Ring Squadron leader until his return to the United States on 27 January 1919, where he was hailed for leading the Americans with 26 victories.

In June 1929, Rickenbacker took a colonel's commission as a Specialist in the Officers Reserve Corps, but he gave it up at the end of the five-year appointment period as he had often stated he always wanted to be remembered as Captain Eddie. During the early part of World War II, he served as the personal observer of Secretary of War Stimson in a flight over Leningrad. On his return to Washington to report on German war damage of the Russian city, his plane was found downed in the Pacific, but he survived a long ordeal on a raft and was eventually rescued.

After World War II, Rickenbacker returned to auto racing and became president of the Indianapolis Speedway. He moved back into aviation and built up Eastern Air Lines into one of the commercial giants. On the last day of 1963, Captain Eddie retired as Eastern's Board Chairman, more than 40 years after his glory days over France.

Lt Frank Luke Jr.

- Leading US ace at the time of his death on 29 September 1918 and second only to Rickenbacker among US World War I aces.
- Awarded the Medal of Honor for his heroic actions on 29 September 1918.

Frank Luke, nicknamed "The Arizona Balloon Buster," was born in Phoenix, Arizona. He enlisted in the US Army in September 1917, learned to fly, and arrived on the front in France in July 1918 where he was assigned to the 27th Aero Squadron. His exceptional bravery earned for him a reputation for being "wild and reckless" but his fellow pilots soon realized that he possessed that certain element which distinguished a great fighter pilot from the others—complete fearlessness.

In September 1918 Luke began a personal campaign against German observation balloons and airplanes. During a seven-day period, 12–18 September, two days of which he did not fly, he scored 13 confirmed victories, including an amazing five victories (two balloons and three airplanes) on the last day. At sunset on 29 September 1918, Luke took off from an advance aerodrome at Verdun to attack balloons in the area of Dun-sur-Meuse. He never returned.

Following the Armistice, US troops found Luke's grave in the cemetery of the small village of Murvaux, France. According to the local French residents, Luke had been shot in the chest by ground fire while circling Murvaux at low altitude, had landed his Spad, crawled from it in the gathering dusk, and had died while firing his service automatic at German soldiers who were searching for him along a creek bank. His remains were removed to the US Military Cemetery at Romagne, France, for permanent burial.

For his last flight on 29 September, Luke was awarded the Medal of Honor for shooting down three enemy balloons while under heavy fire both from the ground and from eight pursuing enemy airplanes, and for strafing enemy troops, killing six of them. The citation on Luke's Medal of Honor reads, "After having previously destroyed a number of enemy aircraft within 17 days he voluntarily started on a patrol after German observation balloons. Though pursued by 8 German planes which were protecting the enemy balloon line, he unhesitatingly attacked and shot down in flames 3 German balloons, being himself under heavy fire from ground batteries and the hostile planes. Severely wounded, he descended to within 50 meters of the ground, and flying at this low altitude near the town of Murvaux opened fire upon enemy troops, killing 6 and wounding as many more. Forced to make a landing and surrounded on all sides by the enemy, who called upon him to surrender, he drew his automatic pistol and defended himself gallantly until he fell dead from a wound in the chest."

At the time of his death, Luke was the leading ace of the US Air Service with 18 confirmed victories, (14 balloons and 4 airplanes).

Maj Gen Mason M. Patrick

- Appointed commander of the Air Service in 1918 and again in 1921 by his West Point roommate, Gen John J. "Black Jack" Pershing.
- Qualified as the world's oldest Junior Military Aviator by earning his wings at the age of 50.
- In 1925, General Patrick was a key instigator urging Congress to create an Air Corps with relations to the Army analogous to that which existed between the Navy and Marine Corps.
- After seeing the US Army Air Corps founded in July 1926, he retired in December 1927 with nearly 40 years of service to the country.

Mason M. Patrick was the first real head of American aviation. Patrick entered West Point in September 1882 and upon graduation stood second in a class of 77. Commissioned a second lieutenant of engineers, his first duty was with the Engineer Battalion at Willets Point, New York. After a couple of instructor tours at West Point, he sailed for Cuba where he commanded the 2d Battalion of Engineers and then as Chief Engineer in the Army of Cuban Occupation until 1909. He served as commander, Engineering School at Washington Barracks until sent to Europe in 1917. In France, he was in charge of engineer instruction for the American Expeditionary Forces.

Although an Army engineer for 30 years, in 1918 General Pershing, Patrick's West Point classmate, appointed him as commander of the Air Service in France. In Pershing's words, there were many fine people in the air arm, but they were "running around in circles," he wanted Patrick to make them go straight. Although knowing virtually nothing about aviation at that point, Patrick was an excellent organizer and administrator. By the end of the war, the Air Service was an efficient and well run combat arm. After the Armistice, Patrick returned to the Corps of Engineers, but in late 1921 he was recalled to the Air Service. His predecessor, Charles Menoher, could not get along with the most famous airman of the day, Brig Gen William "Billy" Mitchell, and in the resulting power struggle, Menoher lost. Because Patrick had managed the difficult airman during the war, he was given the opportunity to do so again. For the next six years Patrick remained at the helm, although Mitchell left the service in 1926. General Patrick retired on 12 December 1927.

While heading the young Air Service, Patrick focused on developing a doctrine that explicitly supported the need for autonomy. Patrick knew the value of airpower, but more importantly, he grasped the limitations and capabilities of airpower at that point in time. Patrick built an airpower agenda that included commercial aviation development, Air Service officer professionalization, the development of airpower doctrine, and legislative initiatives that would set the Air Service on the path to independence.

Lt Gen Frank M. "Andy" Andrews

- First commander of General Headquarters (GHQ) Air Force, appointed by Chief of Staff Gen Douglas MacArthur in 1935.
- Appointed by General Marshall as assistant chief of Staff for Training and Operations (G-3)—the first airman to ever hold a position this high on the Army General Staff.
- Commanding General of US Middle East Forces in 1942.
- Commanding General of all US Forces in newly formed European theater of Operations preceding Gen Dwight D. Eisenhower.
- On 2 May 1943, the B-24 carrying General Andrews crashed against a fog-shrouded promontory while landing at Meeks Field near Keflavik, Iceland.

Frank Andrews was one of the great personalities in Air Force history. He was entrusted with major commands in the Army at a time when airman were often at odds with their Army superiors. Andrews graduated from West Point in 1906 and served his first 11 years as a Cavalry officer in the Philippines, Hawaii, and in the United States. In 1917, he was transferred to the Signal Corps for duty with the Aviation Division and learned to fly in France during World War I.

From August 1920 to February 1923, Andrews commanded the US Army Air Service's European air force of 13 DH-4s under his highly popular father-in-law, Gen Henry Allen. While Andrews was attending the Air Corps Tactical School and the Command and General Staff School at Fort Leavenworth, Kansas, Billy Mitchell was fighting his battles in Washington. Because of this absence, Andrews was never considered one of "Mitchell's Boys."

On 1 March 1935 Andrews became the first commander of GHQ Air Force at Langley Field, Virginia. GHQ Air Force was created to provide centralized command and control of long-range bombardment and observation aircraft not supporting friendly ground forces. He handpicked his staff from the brightest minds in the Air Service, among them Harvey Burwell, Follett Bradley, George C. Kenney, and Joseph T. McNarney. As commander of GHQ

Air Force, Andrews was in a position to translate doctrine into strategy and tactics. He ordered all pilots in GHQ to become instrument rated. Instrument flying enlarged mobility as did ever-extending aircraft range, altitude, and speed. In order to take advantage of long-range doctrine, Andrews oversaw the development of the first truly long-range bomber, the B-17 Flying Fortress.

In August of 1939, Andrews was named assistant Chief of Staff for Training and Operations under Gen George C. Marshall, the first airman to hold this level of position. In September 1941, he was placed in command of the Caribbean Defense Command and the Panama Canal Department—the first unified command. He impressed everyone with his administration of the command and the preparations for war that Andrews had conducted with minimal resources. In early 1943, General Marshall announced the formation of a European theater of operations. General Andrews was selected to command it. In the three short months that Andrews commanded in Europe, he built up a decimated Eighth Air Force and laid the foundation for the buildup that would become Operation Overlord.

General Andrews was killed on 2 May 1943 when his B-24 crashed during a fog-shrouded landing near Keflavik, Iceland.

World War II

1941–45

The Doolittle Raid on Tokyo (18 April 1942)

Aviation pioneer, Gen James H. "Jimmy" Doolittle.

Navy Capt Francis S. Low conceived the idea of flying Army medium bombers off a Navy carrier and attacking Japan. The B-25 was selected because it was small; had sufficient range to carry two thousand-pound bombs, two thousand miles; and because it took off and handled very well. General Arnold selected Lt Col James H. "Jimmy" Doolittle to lead the attack. According to Arnold, "First I found out what B-25 unit had the most experience and then went to that crew, that organization and called for volunteers and the entire group, including the group commander, volunteered."

The training was hard, no one had ever taken off a fully loaded B-25 in less than five hundred feet. First they had to prove it could be done, then they had to train the people to do it. Before they were through, the bombers would lift off in only 287 feet. The crews proved they were good and so were their airplanes.

The raid was carefully planned, nothing was left to chance. Because the attack would be low-level, Norden bombsights were replaced by a twenty-cent improvisation to prevent the secret devices from falling into enemy hands. Doolittle then considered what to do if the Japanese spotted the task force. If intercepted by Japanese surface ships or aircraft, the aircraft would immediately leave the decks. If they were within range of Tokyo they would go ahead and bomb Tokyo, even though they would run out of gasoline shortly thereafter. That was the worst possible scenario. If the aircraft were not in range of Tokyo, they would go back to Midway. If neither Tokyo nor Midway were in range, the B-25s would be pushed overboard so the decks could be cleared for the use of the carrier's own aircraft.

On the morning of 18 April 1942, Japanese patrol boats sighted the task force. The boats were quickly destroyed, but they could have transmitted a position report. It was eight hours before scheduled takeoff, an additional four hundred miles to the target. Gas reserves would be dangerously low, but they were spotted and they would have to go.

The program went almost according to plan. The B-25s were to bomb the targets, turn in a general southerly direction, get out to sea as quickly as possible, and after being out of sight of land, turn and take a westerly course to China.

The bombers came in on the deck and pulled up to about fifteen hundred feet to bomb and to make sure they were not hit by the fragments of friendly bombs. According to Doolittle, the feeling was "Get the job done and get the heck out of there." The actual damage done by the raid was minimal. There were 16 B-25s each carrying one ton of bombs. In later raids, Gen Curtis E. LeMay with his Twentieth Air Force,

Doolittle's B-25 takes off from the carrier *Hornet* en route to Tokyo.

sent out five hundred planes on a mission, each carrying 10 tons of bombs.

Reaching a safe haven after the raid wasn't easy, and because they had to take off much sooner than planned, they were very low on fuel. One crew went to Vladivostok, the other 15 crews proceeded until they got to the coast of China. When they reached China, two of the Mitchell Bombers were so low on fuel that they landed in the surf along side the beach. Two people were drowned, eight of them got ashore. The weather was quite bad, so most of the aircraft flew on until they felt they were as close to their final destination as possible. Having been on dead reckoning for quite awhile, most crews were off target when they jumped.

The crew of Farrow's plane after being captured. In front, L to R, were Cpl Jacob DeShazer and Sgt Harold A. Spatz; in rear, L to R, Lt William G. Farrow, Lt Robert L. Hite, and Lt George Barr. Farrow and Spatz were two of those executed on 15 October 1942.

On 15 August 1942 it was learned from the Swiss Consulate General in Shanghai that eight American flyers were prisoners of the Japanese. After the war, the facts were uncovered in a War Crimes Trial held at Shanghai that opened in February 1946 to try four Japanese officers for mistreatment of the eight prisoners of war (POW) of the Tokyo Raid. Two of the original 10 men, Dieter and Fitzmaurice, died when their B-25 ditched off the coast of China. The other eight, Hallmark, Meder,

Lt Hite, blindfolded by his captors, is led from a Japanese transport plane after he and the other seven flyers were flown from Shanghai to Tokyo. After about 45 days in Japan all eight were taken back to China by ship and imprisoned in Shanghai.

Nielsen, Lt William G. Farrow, Lt Robert L. Hite, Lt George Barr, Sgt Harold A. Spatz, and Cpl Jacob DeShazer were captured. In addition to being tortured, they contracted dysentery and beriberi as a result of the deplorable conditions under which they were confined. On 28 August 1942, Hallmark, Farrow, and Spatz were given a "trial" by Japanese officers, although they were never told the charges against them. On 14 October 1942, Hallmark, Farrow, and Spatz were advised they were to be executed the next day. At 4:30 P.M. on 15 October 1942, the three Americans were brought by truck to Public Cemetery No. 1 outside Shanghai. In accordance with proper ceremonial procedures of the Japanese military, they were then shot.

The other five men remained in military confinement on a starvation diet, their health rapidly deteriorating. In April 1943 they were moved to Nanking and on 1 December 1943, Meder died. The other four men began to receive slight improvement in their treatment and by sheer determination and the comfort they received from a lone copy of the Bible, they survived to August 1945 when they were freed. The four Japanese officers tried for their war crimes against the eight Tokyo Raiders were found guilty. Three were sentenced to hard labor for five years and the fourth to a nine-year sentence.

Eighty crew members flew in the Doolittle Raid, 64 returned to fight again. They were part of a team recognized for its professionalism and heroism, a rich heritage remembered by a new generation of airmen. When the news of the raid was released, American morale zoomed from the depths to which it had plunged following Japan's successes. It also caused the Japanese to transfer back to the home islands fighter units that could have been used against the Allies. In comparison to the B-29 attacks against Japan two years later, the Tokyo Raid was a token effort. However, it was an example of brilliant tactics and achieved a moral victory for the nation.

Kasserine Pass (February 1943)

In 1943, Gen Carl A. Spaatz commanded all Allied air forces in Africa.

Kasserine Pass illustrated the importance of gaining air superiority before undertaking a land battle and the need for centralized control of air assets.

The Battle for Kasserine in central Tunisia, February 1943, was the first serious encounter between Germans and American-led armor forces in World War II. It has been generally regarded as a military embarrassment to Gen Dwight D. Eisenhower, commander of Allied forces in northwest Africa. It laid bare the green, ill-equipped condition of American air and ground warriors. Americans lost thousands (killed, missing, or captured), especially in the first actions in Faid Pass and Sidi-bou-Zid. American "can-do-ism" could not replace proper preparation when tangling with a tough enemy. One of the positive results was that airmen gained greater equality and effective centralized control of air assets as a result of Kasserine battle analysis.

Interestingly, the last of the three phases, this in Kasserine Pass proper, was a victory over the forces of Field Marshal Erwin Rommel. He was forced to pull back his troops, and disappointed, left the African theater, never to return.

American air units—the 33d Fighter Group for one—led by Col William W. "Spike" Momyer (later general), got so worn out in the futile attempt to support the ground forces and try to gain even a modicum of air superiority,

they were forced to retire from action before the battle in Kasserine Pass. Desperate conditions forced General Doolittle to take centralized control of all Twelfth Air Force assets. Bad weather, an old nemesis to effective air support, prevented full application of theater airpower, but Doolittle sent medium and heavy bombers to help turn the tide. In the last days as Rommel pulled back, fighters and bombers hit trucks and tanks, giving a meaningful blow to the enemy armor.

The importance of Kasserine transcended the actual combat, losses or victories. Allied air and ground leaders, vowing never to be pushed around by the Germans in the future, recognized the importance of the long-held doctrinal point of gaining air superiority before undertaking land battle. Leaders in Africa also recognized another historically articulated point of combat organization, that airmen needed to centrally command the air aspect of modern mobile warfare. This was suggested by the recent reformation of the US Army combat capabilities into two equal forces: Army Ground Forces and Army Air Forces (AAF).

US troops inspecting German Tiger tanks destroyed by Allied aircraft in Tunisia.

Battle of the Bismarck Sea (1943)

The B-25 Mitchell Bomber employed skip-bombing techniques in the battle of the Bismarck Sea.

At the same time the Guadalcanal campaign was raging, an equally bitter series of battles was occurring on the island of New Guinea. In early March, a Japanese convoy set out from Rabaul to land much-needed reinforcements in the Buna-Gona area of New Guinea. It was composed of eight transports, escorted by eight destroyers, and screened by an inadequate light combat air patrol.

If the reinforcements were permitted to land, the Japanese would have an almost impenetrable stronghold in New Guinea. It seemed the war in the Pacific theater hinged on the United States stopping this buildup. General MacArthur, Southwest Pacific Area Commander, needed a strategy to stop the convoy and a service to do it. MacArthur could not depend on ground or surface-naval forces because they had been wiped out by previous Japanese attacks so he turned to the Air Force. Under the guidance of Gen George C. Kenney, Fifth Air Force Commander, air units were organized and prepared to engage the convoy.

Unbeknownst to the Japanese, the American air force had been experimenting with a new aerial tactic called skip-bombing, wherein the attacking airplane drops a bomb with a long-delay fuse close to the surface and lets it skip into the side of the target ship. This was the first occasion in which the Americans would use this new tactic. On 3 March as soon as the Japanese came under the radius of American airpower, the convoy was attacked relentlessly. The first day's attack (by high altitude B-17s) sank two transports and damaged a third. Two destroyers were tasked with rescuing the survivors and making a high speed run to New Guinea to deposit them. This they did, and returned to the plodding convoy before dawn the next day.

Coming within range of American and Australian medium bombers, like the B-25 Mitchell Bomber, proved to be a disaster for the Japanese on 4 March 1943. The convoy was savaged by skip-bombing and strafing. By noon, all six remaining transports and four of the destroyers were sinking or sunk. The remaining four destroyers recovered what few survivors they could and fled north to Rabaul. After this, the Japanese would never again attempt to run slow transports into the face of American airpower.

The results of the attack on the convoy in the Bismarck Sea were astounding. The victory changed the direction of the war in the region, preventing what could have been a series of bloody battles of attrition in Southwest New Guinea.

This small Japanese ship is about to be sunk by the 500-lb bomb in the picture.

Ploesti Raid (August 1943)

The B-24 Liberator could carry 5,000 pounds of bombs 2,300 miles.

The oil refineries at Ploesti, Romania, provided Germany with 35 percent of its oil. Air planners figured it would take many high-level attacks by huge fleets of heavy bombers to destroy the refineries. Col Jacob E. Smart, a member of Arnold's Advisory Council, believed a low-level attack might prove successful. Smart had seen A-20s in training hit moving targets while flying low and fast. This led him to conclude that aircraft accuracy would allow a low-level attack of Ploesti. Smart believed a low-level attack might mitigate the extensive air defenses. Not everyone held this opinion. Col Richard Hughes, an AAF target expert, protested that Allied pilots did not have the necessary skills or experience to tackle such a complex mission. However, with President Roosevelt and General Arnold's backing, the mission plans were built.

The plan called for a 2,700-mile mission conducted by more than three hundred bombers to attack seven refineries. The mission was flown by IX Bomber Command whose training included bombing a full-scale outline of the Ploesti complex in the Libyan desert. The operation included the 44th, 93d, 98th, 376th, and 389th Bomb Groups. Because of other operational requirements the units could not focus their training on Ploesti until after

Low-level B-24 Liberators over Ploesti.

19 July 1943, only 11 days prior to the mission date.

On 1 August, 165 B-24s took off to bomb Ploesti. The 376th Bomb Group, commanded by Col Keith Compton, led the mission. The lead navigator was in another plane that ditched into the Mediterranean Sea several hours after launching. Colonel Compton misidentified the initial point, a ground feature used to coordinate the attack, and led his group on a route 30 miles south of Ploesti. Three of the five groups were behind schedule and Compton's error eroded any remaining attack coordination.

The Ploesti raids illustrated the danger of low-level strategic bombing but also the courage and determination of US aircrews.

One hundred sixty-four B-24 Liberators reached Ploesti and attacked at levels often lower than the refineries' towers. Bombers flying through the explosions of oil tanks were assaulted by merciless flak trains and machine gun fire. B-24 gunners dueled with gunners in towers and church steeples. Ploesti was also defended by 120 German and two hundred Romanian fighter aircraft. The 98th Bomb Group, led by Col John Kane, was the only one to fly its assigned course and arrive on schedule. The courage and determination of the aircrews is the sole reason the raid had any success.

Flying so low that aircraft had to ascend to avoid smokestacks 210 feet high, the bombers took high losses. Seventy-three B-24s were lost in the raid and another 55 suffered major damage. Nearly five hundred aircrew were killed or wounded and more than one hundred became prisoners of war. Navigation and timing problems prevented a coordinated attack. Despite this, aircrews managed to destroy 60 percent of the complex's output. Five airmen, including Colonel Kane, received the Medal of Honor for their bravery during the raid—the most of any single engagement in World War II.

The Regensburg/Schweinfurt Raids (1943)

Bomber formations were devised to maximize the number of defensive guns brought to bear against enemy fighters.

The growing strength of German fighter operations in Europe was a great concern to the Allies in 1943. On 10 June 1943 the combined Chiefs of Staff issued the directive known as "Pointblank." This directive placed German fighter strength as the top strategic priority. In order to hurt the German's fighter operation, Colonel Hughes, an Allied air planner, decided to attack production facilities at Regensburg and Schweinfurt. A large percentage of Germany's fighters were produced in Regensburg in southeast Germany. An equally critical target was Schweinfurt, a major ball-bearing production center.

The plan was for General LeMay's 4th Bomb Wing to fly to Regensburg, bomb the Messerschmitt plant, then fly across the Mediterranean and land in North Africa. The arrival of the new B-17F with greater range made this possible. The Luftwaffe was expected to meet the attack early, then land and refuel for the attack on the bombers as they headed back to England. The Allied plan, however, called for the 1st Bomb Wing to follow the 4th by only 15 minutes along the same flight path before breaking off to bomb Schweinfurt. By the time the Germans figured out that the 4th Bomb Wing was not returning and that the 1st was heading for Schweinfurt, they would be on the ground short of both fuel and ammunition. The plan called for the biggest aerial diversion ever attempted with three B-26 groups raiding coastal airfields to draw Luftwaffe fighters from the 1st Bomb Wing. This would allow the 1st to attack Schweinfurt relatively unmolested.

On 17 August 1943, 139 B-17s, with LeMay in the lead, crossed the Dutch coast headed for southeast Germany. The 4th Bomb Wing lost 17 aircraft en route to Regensburg but the remaining 122 bombers conducted a very accurate mission from less than 20,000 feet. As the 4th left Regensburg, the 1st Bomb Wing was still over the North Sea, five hours behind schedule—the timing plan was awry.

The Luftwaffe expected the 4th Bomb Wing to return to England and massed their fighters in unprecedented numbers. The 1st Bomb Wing flew into this mass of three hundred enemy fighters over Germany. By the time the 1st reached the Schweinfurt initial point, 24 B-17s were lost leaving 196 available for bombing. The four ball-bearing plants at Schweinfurt were tough targets to find under ideal circumstances. However, the delay in launching caused an approach heading change to avoid flying into a setting sun. This change combined with the Germans' artificial fog generators made the task nearly impossible. The 10 bomb groups scattered their bombs all over the town while the Luftwaffe refueled and rearmed their aircraft. The return trip for the 1st Bomb Wing was tougher than the trip into Germany.

The Regensburg/Schweinfurt raids cost the Allies 60 B-17s, 16 percent of the dispatched force. General Arnold reported the operation a success. In terms of lost production, the attack on Regensburg probably accounted for one thousand lost Me-109s or about three weeks of total fighter production. At Schweinfurt, the attack proved to have little if any effect. Three of the five factories were severely damaged but few of the machine tools that produced ball bearings were destroyed. "The Raid" showed how difficult and costly it was to conduct air warfare. However, "The Raid" foretold the story of the day when the Luftwaffe would not be able to stop Allied bombing.

"The Big Week" (February 1944)

P-47s were the key to victory during "The Big Week" air battles.

After the heavy losses of the Regensburg/Schweinfurt raids in August of 1943, General Spaatz, commanding the US Strategic Air Forces in Europe decided not to attempt daylight raids deep into Germany until he had assembled an adequate force of long-range fighters that could escort the heavy bombers to their targets. By February 1944, a sizeable number of long-range P-47 "Thunderbolts" were available and P-51 "Mustangs" were starting to arrive. Spaatz knew that his force was ready to get back to its main task: destroying the Luftwaffe.

The leaders wanted an extended period of good weather over Germany (a rare event in the fall and winter) so that they could launch a series of heavy raids. On 20 February, the cloud cover finally broke and Spaatz launched over one thousand heavy bombers, escorted by more than eight hundred fighters. Between 20 and 25 February, US bombers made almost four thousand sorties and dropped 19,000 tons of bombs on key German aircraft and ball-bearing factories. In addition to Regensburg and Schweinfurt, the B-17s and B-24s attacked Leipzig, Stuttgart, Brunswick, Oschersleben, and Gotha. These attacks became known as "The Big Week."

Big Week saw some of the largest and most ferocious air battles in history. By the end of the week, US aircrews were exhausted but the air-war had been transformed. United States losses were heavy: 226 bombers were lost along with 28 fighters and more than twenty-five hundred air crewmen were killed, seriously wounded, or captured. But German losses were much worse. The bombs did severe damage to the German aircraft industry and the US fighters had devastated the Luftwaffe. During the month of February, the Luftwaffe lost 33 percent of its daylight fighters and 20 percent of its fighter pilots, including some of its leading aces. The German tactic of attacking the US bombers and ignoring the US fighters had led to heavy US bomber losses but had left the US fighters free to annihilate the German fighters.

During a week of ferocious air battles, US planes destroyed 33% of the Luftwaffe's fighter planes and killed 20% of its fighter pilots.

After Big Week, the US Strategic Air Force had turned the corner. In a few months, the United States would gain air superiority over all of Europe. The Germans would later change their tactics, and bring in jet fighters, but after Big Week, their efforts were too little and too late. There was another year of hard fighting ahead in Europe, but once the results of Big Week became clear, the outcome of the war was no longer in doubt.

Operation Overlord (1944)

P-38 Lightnings were assigned to provide air cover over the invasion fleet to prevent confusion for Allied gunners.

On 6 June 1944 the Allies invaded occupied Europe. Code named Operation Overlord, the assault hit five beaches stretched along 50 miles of French coastline. The largest amphibious assault ever, it required massive air support for success. By the day's end, the IX Troop Carrier Command dropped 13,000 paratroopers, 4,400 glidermen, and thousands of dummy parachutists on France. The IX Bomber Command attacked the defenses at Utah beach. Many bombers were assigned to bomb the beaches prior to troops landing; the bomb craters would serve as ready-made foxholes. Heavy bombers were used to destroy the towns surrounding Normandy to create roadblocks and prevent resupply.

To hold down friendly-fire casualties, broad white strips were painted around the wings and fuselage of all aircraft. As a further precaution, Gen William Kepner assigned the P-38 Lightning units to fly support over the fleet. The Germans had no aircraft that resembled the twin-fuselage of the P-38. The plan was nearly flawless, the Navy only shot down two American aircraft, both P-51s.

The Allies operated with air dominance over Normandy. The two strategic air forces had thirty-five hundred heavies and the tactical units operated fifteen hundred mediums and fifty-five hundred fighters. By comparison, the Luftwaffe had only 319 aircraft in all of France. Hitler refused to divert fighters to Normandy because he felt the attack was a feint. Air superiority allowed Allied trucks, tanks, and guns to move in broad daylight without interference. In planning Overlord, a German counterattack was projected for D day plus 14. However, it never occurred because air interdiction made enemy troop movement or massing impossible. Nevertheless, the Allied troops were engaged in a stalemate against the German defensive positions—a breakout was needed.

To breakout from Normandy, Eisenhower suggested carpet bombing. One operation, code named Cobra, began on 19 July 1944. The bombing operation planned to saturate a rectangle three miles long and one and one-half miles wide and destroy the Panzer Lehr division. Gen Omar N. Bradley wanted the bombardment to occur only eight hundred yards in front his troops. Eighth Air Force recommended a separation of three thousand yards. Furthermore, Bradley wanted the bombers to fly parallel to the road, through the three-mile length of the rectangle. Gen Hoyt S. Vandenberg advised against this because it would take hours to fly fifteen hundred aircraft through a zone only one and one-half miles wide. Bradley agreed. On 25 July 1944, Cobra was launched. As predicted, with only an eight hundred yard buffer, short bombs fell on the 30th Infantry Division killing or wounding five hundred soldiers including Lt Gen Leslie J. McNair. Despite these setbacks, the carpet bombing was successful. When the VII Corps attacked it met only sparse resistance. By 31 July, American troops were on the advance.

Troops deploy on Normandy's beaches.

Bombing Tokyo (9–10 March 1945)

The B-29 was the best bomber of the war and the main weapon in the strategic bombing of Japan.

The Doolittle raid on Tokyo had done a lot for US morale, but it was not a viable method of conducting sustained strategic bombing of Japan. To bring the full weight of airpower to bear on Japan, the United States built bombers of unprecedented range (the B-29s) and captured islands from which B-29s could reach Japan. In November 1944, US bombers finally returned to Tokyo in the form of B-29s flying out of Saipan.

The Japanese tried to defend against the B-29s but they grossly underestimated the power of strategic bombing. They chose to concede air superiority over Japan to the United States in order to use their main air strength to oppose Allied surface forces. This disastrous miscalculation was partly due to the fact that Japanese air forces were tightly controlled by army and navy leaders who had a weak understanding of airpower.

Limited Japanese defenses, however, did not guarantee successful US bombing. The B-29 was the best bomber of World War II, but it did not have the ability to hit precision targets through clouds. The consistently bad weather over Japan made sustained precision bombing of Japanese factories impossible. On the night of 9 March 1945, General LeMay, commanding B-29 operations against Japan, ordered a radical change in tactics. On this raid the B-29s did not execute their normal daylight, high-altitude, formation attack with high-explosives but instead, they hit Tokyo with incendiary bombs, at night, from low altitude, flying individually. Flying at night at low altitude took advantage of Japanese weakness in night-fighters and low-altitude air defenses. The low-altitude individual bombing runs also enabled the B-29s to carry less fuel and hence more bombs. Since LeMay expected little enemy fighter opposition, he removed the gunners, guns, and ammunition from the B-29s and replaced them with still more bombs. The change in tactics doubled the bomb-load of each B-29 and incendiary bombs were much more effective against the highly flammable Japanese city than high-explosive bombs.

The Japanese underestimated air-power and conceded air superiority over Japan to the United States in order to concentrate their forces against a surface invasion.

The results were stunning. Before 9 March 1945, the B-29s had done very little damage to the Japanese war effort, but on that night they burned out 16 square miles of Tokyo and killed more than 80,000 Japanese in the most devastating air raid ever. Subsequent firebombing devastated more than 60 Japanese cities, left millions of Japanese homeless, and radically reduced Japan's military production.

Gen LeMay ordered a radical change in tactics.

The Hiroshima and Nagasaki Bombings
(August 1945)

The modified B-29, *Enola Gay*.

President Truman and the armed forces had three strategic options for inducing the Japanese surrender: continue the fire-bombing and blockade, invasion, or use the atomic bomb. Truman was aware that the first two options would probably not be very effective methods to induce the Japanese to surrender. The Battle for Okinawa caused 48,000 American casualties when the Japanese refused to surrender. So it was the right time to resolutely make a decision. Gradually, US authorities made preparations for the decision to use the bomb, as it was close to production. The Interim Committee on S-1 suggested to the president that the bomb should be used directly against Japan, because a demonstration explosion was not thought to be a strong enough representation of the power that the bomb held. Several US military leaders went with the president to the Big Three meeting at Potsdam in July, and discussions continued there. It was determined then that the bomb should be used. On 25 July Truman prepared the order for use of the first atomic bomb as soon after 3 August as weather permitted on one of the four target cities. The Potsdam Proclamation was issued during the Potsdam meeting by the heads of government of the United States, Britain, and China. It warned of "utter devastation of the Japanese homeland" unless Japan surrendered unconditionally.

At approximately 2:00 A.M. on the morning of 6 August, the *Enola Gay*, carrying an atomic bomb, started on the long flight from Tinian Island. The *Enola Gay* was one of 15 B-29s modified specifically for the highly secret atomic bomb missions. These airplanes were outfitted with new engines and propellers and faster-acting pneumatic bomb bay doors. Two observation planes carrying cameras and scientific instruments followed behind her. After 6:00 A.M., the bomb was fully armed on board the *Enola Gay*. Col Paul W. Tibbets Jr., pilot of the *Enola Gay*, announced to the crew that the plane was carrying the world's first atomic bomb. The trip to Japan was smooth. At about 7:00 A.M., the Japanese radar net detected aircraft heading toward Japan, and they broadcast the alert throughout the Hiroshima area. Soon afterward when an American weather plane circled over the city, the people went back to their daily work thinking the danger had passed. At 8:00 A.M. the Japanese detected two B-29's heading toward Hiroshima. The radio stations quickly broadcast a warning for the people to take shelter, but many did not follow the advice. They thought that it was the same as the first time. At 8:09 A.M., the crew of the *Enola Gay* at 26,000 feet could see the city appear below; it was time to drop the bomb. Just then, they received a message indicating that the weather was good over Hiroshima. The bomb was released at 8:16 A.M. A terrible, strong, and unimaginable explosion occurred near the central section of the city. The crew of the *Enola Gay* saw a column of smoke rising fast and intense fires springing up. The astonishing result of the first atomic strike: it killed between 70,000 and 80,000 people, injured another 70,000, and burned almost 4.4 square miles. On 9 August, Nagasaki was bombed by a B-29 named *Bock's Car*. The Japanese unconditionally surrendered on 14 August 1945.

General of the Army Henry H. "Hap" Arnold

- Taught to fly by Orville and Wilbur Wright.
- A five-star general and two-time winner of the Mackay Trophy for aeronautical achievement.
- In 1934, he was awarded the Distinguished Flying Cross for demonstrating the range of strategic airpower by leading a flight of B-10 bombers from Washington, D.C., to Alaska.
- Named Chief of the Army Air Corps in 1938. During World War II, he became the first Air Chief to sit as an equal member on the Joint Chiefs of Staff.
- He was the commanding general of the Army Air Forces (AAF) during World War II.

Henry H. Arnold was one of the truly great men in American airpower. Taught to fly by the Wright brothers, he rose steadily in rank and responsibility throughout the '20s and '30s and became the commanding general of the Army Air Forces during World War II. In 1944 he was promoted to five-star rank, but his health was very poor, he suffered several heart attacks during the war, and retired less than a year after Japan surrendered.

Graduating from West Point in 1907, Arnold had hoped to join the cavalry. However, his cadet performance was so dismal he was relegated to the infantry. After a tour in the Philippines, he reapplied to the cavalry, but again was refused. Largely out of a desire to escape from the infantry, Arnold then applied for the Signal Corps and became one of America's first military pilots. Aviation was extremely dangerous in those early days, and after several crashes and near crashes, Arnold elected to ground himself. After more than three years of desk work, he overcame his fears and returned to flying. Because of his relatively extensive experience in aviation, and much to his chagrin, he was forced to remain in Washington on the Air Service staff during World War I.

After Armistice Day, he slowly began his steady rise in rank and responsibility. He commanded wings and bases, became a protégé of Billy Mitchell, twice won the Mackay Trophy for aeronautical achievement, was awarded the Distinguished Flying Cross for leading a flight of B-10 bombers to Alaska to display the range of strategic airpower, and was named assistant to the chief of the Air Corps in 1935. When Oscar Westover was killed in a plane crash in 1938, Arnold succeeded him as chief. In this position he was instrumental in laying the groundwork for the massive industrial expansion the war would require. During the war he sat as an equal member of the Joint Chiefs of Staff and was responsible for guiding the air strategy of the various theaters. Belying his nickname "Hap" (short for "happy"), Arnold was a difficult taskmaster. He drove himself so hard during the war that he suffered several heart attacks and he pushed his subordinates just as hard. This did not endear him to everyone in the USAAF, but it was just what was needed to run the largest air force during the largest war in the history of the world. His drive, vision, and sense of initiative were indispensable in leading the air arm during the war and setting the stage for the creation of the US Air Force shortly after the war.

Lt Gen Claire L. Chennault

- In the 1930s, he was the Air Corps Tactical School's most famous proponent of pursuit tactics at a time when strategic bombardment was premier.
- Forced out of the Air Corps in 1937 because of bronchitis, he went to China to advise Chiang Kai-shek on building an air force.
- Commanded the American Volunteer Group, better known as the "Flying Tigers."
- Under his leadership the Flying Tigers overcame severe operational handicaps and achieved a two-to-one kill ratio over the Japanese.

Claire L. Chennault's reputation as leader of the Flying Tigers has been immortalized in movies and novels, making him one of America's more famous airmen. Chennault arrived at the Air Corps Tactical School (ACTS) in 1930 with a reputation as a premier pursuit pilot. His ideas concerning pursuit employment evolved from much thought and practical experience. But Air Corps doctrine was making a decisive shift in favor of bombardment, and Chennault's attempts to stem that tide were futile. Chennault's abrasive personality negated his arguments, and his colleagues found it more satisfying simply to ignore him. Suffering from a variety of physical ailments and realizing his theories were out of tune with Air Corps policy, he retired in 1937. Soon after, he traveled to China, where he served as an adviser to Chiang Kai-shek, and formed the Flying Tigers volunteer group to fight against the Japanese. The much-storied group of mercenaries-turned-heroes was well suited to Chennault's aggressive and unconventional personality.

When America entered the war, the Flying Tigers were incorporated into the Fourteenth Air Force, and Chennault was promoted to brigadier general and made its commander. Chennault was an outstanding tactician, whose determination in the face of overwhelming supply and equipment difficulties kept the Fourteenth Air Force in the field, but his strategic ideas were limited to his tactical mindset. Never on good terms with his Air Corps colleagues, Chennault exacerbated this relationship with his constant complaints and his tendency to circumvent the chain of command by dealing directly with Chiang and President Roosevelt. Although knowing how this infuriated his superiors, Chennault persisted. As a consequence, George Marshall thought him disloyal and unreliable. Hap Arnold and Joe Stilwell disapproved of his command style. Even if his strategic theories had been correct, his method of promoting them ensured their demise. He believed that a small force of aircraft, mostly pursuit with a handful of bombers, could so disrupt Japanese logistics as to lead to its eventual defeat. In retrospect, it is doubtful if any amount of tactical airpower could have prevented Japan from overrunning China, much less brought about its defeat.

Lt Gen Ira C. Eaker

- Lt Gen Ira Eaker was commander of the VIII Bomber Command in England which became the Eighth Air Force in 1944.
- He piloted the *Question Mark* in its record-breaking air refueling flight over California in 1929. The plane remained aloft for 150 hours, 40 minutes, and 15 seconds.
- Served as aide to Maj Gen James Fechet, the Air Corps Chief, and as private pilot to Maj Gen Douglas MacArthur.
- In 1927 he piloted the *San Francisco*, the only plane to complete a 23,000 mile Pan American goodwill flight on schedule. For this he was awarded his first Distinguished Flying Cross.
- During World War II, he commanded Allied Air Forces in the Mediterranian.
- He was deputy commander of the Army Air Forces in 1945–46.

One of the great pioneer airmen, Ira C. Eaker, met "Hap" Arnold and Carl Spaatz at Rockwell Field in 1918, and the three became friends and colleagues for life. Eaker was one of the premier pilots between the wars, participating in the Pan American flight of 1926–27 and the *Question Mark* flight of 1929. The *Question Mark* project was the product of Eaker's imagination, political savvy, and zeal. He selected a trimotored Fokker and a Douglas C-1 for the flights. On 1 January 1929, the Fokker took off from San Diego, California, and droned back and forth at 70 MPH between Los Angeles, California, and San Diego for six days. Eaker piloted the *Question Mark* with Pete Quesada as copilot and Maj Carl Spaatz in back to hook up the hose during refueling. On 7 January the Fokker's left engine quit and the *Question Mark* was forced to land with a record-breaking 150 hours, 40 minutes, and 15 seconds aloft.

Eaker was also politically well connected, serving not only as an aide to Maj Gen James Fechet, the Air Corps chief, but also as the private pilot of Gen Douglas MacArthur. An excellent writer with a graduate degree in journalism, he figured prominently in airpower public relations efforts during the 1930s and coauthored several aviation books with Hap Arnold. During World War II he joined Spaatz in England to head the VIII Bomber Command and eventually Eighth Air Force. In early 1944 Eaker moved down to Italy to command the Mediterranean Allied Air Forces.

The task of organizing and standing up the Eighth was extremely daunting. Eaker's talents as a leader and manager were essential. Strategic bombing was not a proven concept, the green Eighth was entering combat against an enemy already battle tested, and the prodigious production capacity of America not yet manifest. Moreover, just as it appeared the Eighth was strong enough to play a major role in the war against Germany, it was stripped of men and machines for operations in North Africa and then Italy. Arnold pushed Eaker to do more, and finally, against Eaker's wishes, he was promoted and moved to Italy, while his place at Eighth was taken by James H. "Jimmy" Doolittle. Soon after Doolittle took over, Eaker's labors bore fruit: air superiority over the Luftwaffe was gained, the invasion of France took place, and the sweep across northern Europe eventually led to victory.

In April 1945, Eaker was named deputy commander of the AAF and chief of the Air Staff. He retired from active duty on 31 July 1947.

Gen Carl A. "Tooey" Spaatz

- One of the most favored American air commanders of World War II. Both Generals Eisenhower and Bradley rated Spaatz the best combat leader in the European theater.
- Received the Distinguished Service Cross for shooting down three German aircraft during World War I.
- The project leader for the *Question Mark* flight which refueled in the air to stay aloft over 150 hours. Spaatz rode in the rear of the aircraft where he reeled in and hooked up the refueling hose from the tanker plane.

Carl A. Spaatz was born 28 June 1891, in Boyertown, Pennsylvania. In 1910 he was appointed to the United States Military Academy. Upon graduation on 12 June 1914, he was commissioned into the Infantry. He served with the Twenty-fifth United States Infantry at Schofield Barracks, Hawaii, from 4 October 1914 to 13 October 1915, when he was detailed as a student in the Aviation School at San Diego, California, until 15 May 1916.

Spaatz went to France with the American Expeditionary Forces in command of the 31st Aero Squadron and joined the 2d Pursuit Group in September 1918. He was officially credited with shooting down three German Fokker planes, and received the Distinguished Service Cross. After World War I he reverted to his permanent rank of captain, 27 February 1920, but was promoted to major on 1 July 1920.

Spaatz commanded the Army plane *Question Mark* in its refueling endurance flight over southern California, 1–7 January 1929, keeping the plane aloft a record total of 150 hours, 40 minutes, and 15 seconds, and was awarded the Distinguished Flying Cross.

A few weeks after Pearl Harbor, in January 1942, General Spaatz was assigned as chief of the AAF Combat Command in Washington, D.C. In May 1942 he became commander of the Eighth Air Force to prepare for the American bombing of Germany. On 7 July he was appointed commanding general of the AAF in the European theater an addition to his duties as commander of Eighth Air Force.

On 1 December 1942, Spaatz became commanding general of the Twelfth Air Force in North Africa. He returned to England in January 1944, to command the US Strategic Air Forces in Europe, which he headed throughout the preinvasion period and the ensuing campaign which culminated with the utter defeat of Germany. His service in Africa won an award of the Distinguished Service Medal, and the accomplishments of his Strategic Air Force in 1944 earned him the Robert J. Collier Trophy for that year, awarded annually to the American making the most outstanding contribution to aviation. He was present at all three signings of unconditional surrender by the enemy—Rheims, Berlin, and Tokyo. In February 1946, he was nominated to become commander of the Army Air Forces. In September 1947 he was appointed by President Harry S. Truman as the First Chief of Staff of the new United States Air Force until 30 April 1948. General Spaatz retired on 30 June 1948.

Col Francis S. "Gabby" Gabreski

Francis S. Gabreski was the leading American ace in the World War II European theater with 28 victories. He was born of Polish immigrants in Oil City, Pennsylvania, and grew up bilingual. While he worked his way through college at Notre Dame, he became interested in airplanes and scraped together money for flying lessons. He did not excel. After his second year in college he joined the Army Air Forces and squeaked through primary pilot training. His first assignment was at Wheeler Field, Hawaii. On 7 December 1941 Gabby's was one of the first aircraft launched to chase the Japanese armada. Taking off in an obsolete P-36, he was too late to catch the departing enemy, however, he was shot at by his own nervous troops upon his return. After many delays, Gabby was posted to the Polish-manned 315th Squadron at Northolt in the United Kingdom. After gaining much needed experience flying in combat, Gabby was assigned to the 56th Fighter Group flying P-47 Thunderbolts. On 24 August 1943 he scored his first victory. From that day on, victories came frequently, often by doubles and triples. After flying 193 missions, the AAF ordered him home. While waiting for the transport, Gabby learned of a mission scheduled for that morning. He convinced the commander to let him fly just one more. While strafing an enemy airfield, his prop hit a rise at the end of the field and he was forced to belly-in. After eluding the enemy for five days, he was captured and spent the remaining eight months of the war as a POW in Stalag Luft 1 in Germany.

During the Korean War, Gabreski commanded the 51st Fighter Wing. He played a major role in developing fighter tactics in the newly emerging jet war. Gabby shot down six and one-half MiG-15s between July 1951 and April 1952. He is one of only seven USAF pilots who were aces in both World War II and the Korean War.

Capt Colin P. Kelly Jr.

Capt Colin P. Kelly Jr. was America's first hero of World War II. He was credited at that time after attacking and sinking the Japanese battleship *Haruna* off Luzon in the Philippines on 10 December 1941. While returning to base, Kelly's B-17 was set on fire by Japanese airplanes and one waist gunner was killed. Kelley then ordered the remainder of the crew to bail out. All were saved but Kelly who died when the B-17 crashed to the ground. President Franklin D. Roosevelt posthumously conferred the Distinguished Service Cross upon Kelly for his sacrifice. It was not until after World War II had ended that it was learned that at that early period of the war, the *Haruna* was operating hundreds of miles away off Malaysia. Kelly's airplane had attacked a light cruiser named *Ashigara* and not the *Haruna*.

Gen Laurence S. Kuter

- One of the more accomplished air planners and staff officers in Air Force history.
- On the faculty of the Air Corps Tactical School from 1935 to 1939.
- In 1933, flew alternate wing position with Capt Claire Chennault's acrobatic group, "The Flying Trapeze." This was the first recognized aerial acrobatic team in military service.
- As a major, he was one of the four principal authors of the basic plan for employment of airpower in World War II.

Laurence S. Kuter was born in Rockford, Illinois, on 28 May 1905. He graduated from West Point on 14 June 1927. Originally an artillery officer, serving at the Presidio of Monterey, California, he saw a need for better air observation. Smitten by flight, he asked for a transfer to the Air Corps. He graduated from pilot training in June 1930. In 1933 Lieutenant Kuter flew alternate wing position with Capt Claire Chennault's acrobatic group, The Flying Trapeze. This was the first recognized aerial acrobatic team in military service.

One of the more accomplished air planners and staff officers in Air Force history, he was on the faculty of the Air Corps Tactical School from 1935 to 1939. Kuter was a staunch strategic bombardment advocate. In 1941, he was one of four officers tasked to write Air War Plans Division (AWPD)1, *Munitions Requirements of the Army Air Force,* the seminal war plan that served as the blueprint for the air assault on Germany. Promotion followed quickly. The youngest general officer in the Army in 1942, he served on General Marshall's War Department staff and Arnold's AAF staff; commanded a bomb wing in England; was deputy commander of the Northwest African Tactical Air Force; then returned to Washington for the rest of the war, even representing the AAF at the Yalta conference in 1945 during Arnold's illness. After the war he again served on the Air Staff, headed the Military Air Transport Service during the Berlin airlift, commanded Air University (then Pacific Air Forces), and completed his career as a full general and commander of North American Air Defense Command (NORAD).

Because of Hap Arnold's illness (one of his five heart attacks), then Major General Kuter was designated to attend the Allied conference at Yalta in February 1945 as the AAF representative. The Americans had hoped to establish a communication system to coordinate the air efforts of the three countries to avoid the danger of fratricide. In addition, the United States pushed for an agreement to site B-29 bases near Vladivostok, USSR, from which to bomb Japan. With the end of the war in Europe approaching, however, the Soviets had little incentive to be agreeable: they rejected both proposals.

Maj Richard I. "Dick" Bong

- America's leading ace of all wars.
- Credited with 40 aerial victories in World War II.
- Awarded the Medal of Honor in 1944 personally by Gen Douglas MacArthur in the Pacific, who lauded him as the greatest fighter ace of all Americans.
- Died test piloting a P-80 at Muroc, California, in August 1945.

After schooling in his hometown, Dick Bong enlisted as a flying cadet at nearby Wausau, Wisconsin, 29 May 1941. He took flying training at Tulare and Gardner Fields, California, and Luke Field, Arizona, receiving his wings and commission on 9 January 1942. He instructed other pilots at Luke until May when he went to Hamilton Field, California, for combat training in P-38s.

He went to the Pacific as a fighter pilot with the 9th Fighter Squadron of the 49th Fighter Group in Australia. In November 1942, Major Bong was reassigned to the 39th Squadron of the 35th Group with which he destroyed five Japanese fighter planes before returning to the 9th Squadron in January 1943. He flew with the 9th until November, being promoted to first lieutenant in April and to captain in August.

On 11 November 1943 he was given 60 days of leave and reassigned to Headquarters V Fighter Command, New Guinea, as assistant operations officer in charge of replacement airplanes. In this assignment Major Bong continued to fly combat missions in P-38s and increased his enemy aircraft kills to 28. In April 1944 he was promoted to major and sent home to instruct others in the art of aerial superiority, with assignment to Foster Field, Texas.

In September 1944 he returned to the Pacific with the V Fighter Command as gunnery training officer. Though not required to perform further combat flying, he voluntarily put in 30 more combat missions over Borneo and the Philippine Islands, destroying 12 more planes to bring his total to 40.

Gen George C. Kenney, his overall commanding officer, who later wrote Major Bong's biography, decided that two hundred missions and five hundred combat hours were enough for any individual and ordered him returned to the United States, in December 1944, with recommendation for the Medal of Honor for his second overseas tour. Major Bong then became a test pilot at Wright Field, Ohio. In June 1945 he went to Burbank, California, as chief of Flight Operations and Air Force Plant Representative to Lockheed Aircraft Company, then engaged in developing and manufacturing P-80 jet aircraft.

Major Bong received a full training course for P-80s at Muroc Lake Flight Test Base, California, but died that August when his plane's engine failed during a flight over Burbank.

In combat he had earned the Distinguished Service Cross, two Silver Stars, seven Distinguished Flying Crosses, and 15 Air Medals, in addition to the Medal of Honor.

Lt Gen James H. "Jimmy" Doolittle

- In 1921 Doolittle led a flight of DH-4s in Mitchell's 1st Provisional Air Brigade in the scientific sinking of German battleships.
- He received one of the first PhDs awarded in aeronautical engineering from Massachusetts Institute of Technology (MIT) and established himself as one of the finest racing pilots in the United States before World War II.
- In April 1942 he led 16 B-25s from the deck of the carrier *Hornet* and scored a stunning coup by bombing Tokyo.
- During World War II, Gen Jimmy Doolittle commanded the Twelfth, Fifteenth, and Eighth Air Forces.

James H. "Jimmy" Doolittle was educated in Nome, Alaska; Los Angeles Junior College, California; and the University of California School of Mines, California. He enlisted as a flying cadet in the Signal Corps Reserve in October 1917, and trained at the School of Military Aeronautics, University of California, and Rockwell Field, California. He was commissioned a second lieutenant in the Signal Corps Aviation Section on 11 March 1918.

In September 1922 he made the first of many pioneering flights which earned him major air trophies and international fame. He flew a DH-4, equipped with crude navigational instruments, in the first cross-country flight, from Pablo Beach, Florida, to San Diego, California, in 21 hours and 19 minutes. He made only one refueling stop, at Kelly Field, Texas. The military gave him the Distinguished Flying Cross for this historic feat.

He won the Schneider Cup Race, the World Series of seaplane racing in 1925, with an average speed of 232 MPH in a Curtiss Navy racer equipped with pontoons. This was the fastest a seaplane had ever flown, and Doolittle received the Mackay Trophy for this feat.

At Mitchel Field, in September 1928, he assisted in the development of fog flying equipment. He helped develop the now almost universally used artificial horizontal and directional gyroscopes and made the first flight completely by instruments. He attracted wide newspaper attention with this feat of blind flying and later received the Harmon Trophy for his experiments.

He was promoted to lieutenant colonel 2 January 1942, and went to Headquarters, Army Air Force to plan the first aerial raid on the Japanese homeland. He volunteered and received Gen H. H. Arnold's approval to lead the attack of 16 B-25 medium bombers from the aircraft carrier *Hornet* with targets in Tokyo, Kobe, Osaka, and Nagoya. The daring one-way mission on 18 April 1942 electrified the world and gave America's war hopes a terrific lift. Colonel Doolittle received the Medal of Honor, presented to him by President Roosevelt at the White House, for planning and leading this successful operation.

In July 1942, as a brigadier general, he had been advanced two grades the day after the Tokyo attack, Doolittle was assigned to the Eighth Air Force and in September became commander of Twelfth Air Force in North Africa. He was promoted to major general in November, and in March 1943 became commander of the North African Strategic Air Forces. He took command of Fifteenth Air Force in November, and from January 1944 to September 1945, he commanded Eighth Air Force in Europe and the Pacific, until war's end, as a lieutenant general, the promotion date being 13 March 1944.

Gen George C. Kenney

- George C. Kenney was a fighter pilot during World War I. He downed two German aircraft and won the Distinguished Service Cross.
- Commander of Fifth Air Force and Far East Air Forces providing airpower for Gen Douglas MacArthur in the Southwest Pacific Theater during World War II.
- One of only four airmen to hold the rank of four-star general during World War II.
- One of the most innovative operational airmen of World War II.
- The first commander in chief of Strategic Air Command from 1946 to 1948.

George C. Kenney was America's top airman in the Pacific theater during World War II. Kenney had served as a fighter pilot in the First World War, downing two German aircraft and winning the Distinguished Service Cross. Between the wars he attended Command and General Staff College, the Army War College, and taught at the Air Corps Tactical School before heading Operations and Training for General Headquarters Air Forces. He also earned a reputation as an accomplished engineer through assignments at Wright Field, and became recognized as an expert in tactical aviation. Significantly, he was serving as an air attaché to Paris during the German invasion of France in 1940 and witnessed the effectiveness of airpower in that campaign.

In July 1942, Arnold selected Kenney to become Douglas MacArthur's air deputy. For the rest of the war the short, fiery, and tireless Kenney served as commander of the Fifth Air Force and then Far East Air Forces under the difficult and demanding MacArthur. His success in such battles as Bismarck Sea, Rabaul, Wewak, and the Philippine campaign were dramatic, and he has become the prototype for

the modern concept of an "air component commander," the one individual in charge of all aviation assets in a theater. Kenney's grasp of what is today called "operational art" and how airpower could be used to complement the operations of land and sea forces was outstanding, and he was considered by many to be the most accomplished combat air strategist of the war.

In April 1945 he was promoted to full general— one of only four airmen holding that rank during the war. However, Arnold had more complete confidence in Spaatz and after the war Spaatz was named Arnold's successor. Kenney had hoped to become Chief of Staff after Spaatz but Hoyt Vandenberg, nine years younger than Kenney, replaced Spaatz as chief of staff in 1948. Kenney was instead given command of the new Strategic Air Command (SAC) after the war. When the Berlin Crisis of 1948 broke out, Vandenberg conducted an investigation of SAC's war readiness. The results were unacceptable, so Vandenberg replaced Kenney with Curtis E. LeMay. Kenney was then named commander of Air University. He retired from that position in 1951.

Lt Gen Ennis C. Whitehead

- Participated in Billy Mitchell's bombing tests against the *Ostfriesland* in 1921.
- Joined the Pan American flight of 1927—where he narrowly escaped death in a midair collision over Buenos Aires.
- Set a speed record from Miami, Florida, to Panama in 1931.
- George Kenney's strong right arm in the Pacific, he became commander of the Fifth Air Force in 1944.
- Responsible for such innovations as skip-bombing, parafrag bombs, nose cannons in medium bombers, and the use of combat airlift.

Ennis C. Whitehead is another of the largely forgotten figures of American airpower, although he played an important role at an important time. Enlisting in the Army in 1917, Whitehead quickly joined the Air Service, won his wings, and was sent to France. He was an excellent pilot, but as a result he was made a test pilot and thus saw no combat. After the war, his reputation as an aviator grew within the small coterie of military airmen. He participated in Billy Mitchell's bombing tests against the *Ostfriesland* in 1921, joined the Pan American flight of 1927, where he narrowly escaped death in a midair collision over Buenos Aires, and set a speed record from Miami, Florida, to Panama in 1931. When war came, he was sent to the Pacific where he became George Kenney's strong right arm. Whitehead stayed in Asia for the next seven years, becoming commander of the Fifth Air Force in 1944; and after Kenney left the theater, he took over the Far East Air Forces. Returning to the States in 1949, Whitehead commanded the shortlived Continental Air Command and then the Air Defense Command until his retirement in 1951.

In addition, although Kenney has received credit for such innovations as skip-bombing, parafrag bombs, nose cannons in medium bombers, and the use of combat airlift, it was Whitehead who actually pioneered them. Unquestionably, Whitehead was an outstanding tactician who performed extremely well in the Southwest Pacific theater. Whitehead was a hard, uncompromising man with a heavy twinge of chauvinism; he was a good combat commander who engendered respect rather than admiration among his subordinates. He was also seen by some in the Air Force hierarchy as too attached to Kenney and MacArthur, too political, too outspoken, and too tactically focused. He was disappointed by President Trumon's choice of Hoyt Vandenberg rather than Kenney as Chief of Staff in 1948 and was outraged when the new chief quickly relieved Kenney as commander of Strategic Air Command. These feelings, combined with ill health, caused him to teder his resignation in early 1951.

An MB-2 attempts to sink the USS *Alabama* during Mitchell's 1921 bombing trials.

Lt Gen Benjamin O. Davis Jr.

- The first black to graduate from West Point this century and later became the first African-American Air Force general.
- During his years at West Point he was officially "silenced" by all cadets—no one spoke to him for four years except on official business.
- Commissioned in 1936, earned his wings at Tuskegee in 1941 and was a lieutenant colonel squadron commander in August 1942.
- Commanded the all-black 99th Fighter Squadron in North Africa in 1943 and a fighter wing in Korea in 1953.
- His commands culminated with his third star and command of Thirteenth Air Force.

During World War II, a group of blacks was sent to Tuskegee Institute in Alabama and trained as pilots. The famous Tuskegee airmen went on to serve with distinction in the European theater and the years thereafter. The most famous of these was Benjamin O. Davis Jr.

Davis was the first black to graduate from West Point in this century. His four years there were not, however, pleasant. Because he was black, he was officially "silenced" by all cadets—no one spoke to him for four years except on official business; he roomed alone; he had no friends. That so many cadets, faculty members, and senior officers could allow such behavior is astonishing. This was surely one of the most shameful chapters in West Point history. Nonetheless, Davis graduated but was promptly turned down for pilot training—no black officers were allowed in the Air Corps. While he was serving in the infantry in 1940, this policy was reconsidered, and Davis was sent to Tuskegee for pilot training. Because of the war and his ability, promotion followed rapidly, and soon he was a lieutenant colonel commanding the 99th Fighter Squadron in combat. After one year with this all-black unit in Italy, Davis was promoted to colonel and tasked to form the 322d Group. This black fighter group served admirably for the remainder of the war.

Segregation ended in the services in 1948 with a presidential decree. Davis then attended Air War College, served in the Pentagon, and was sent to Korea in 1953 to command a fighter wing. The following year he received his first star and moved to the Philippines as vice commander of the Thirteenth Air Force. After tours in Taiwan, Germany, the Pentagon, and a return to Korea—while also gaining two more stars—Davis became commander of the Thirteenth Air Force at Clark Air Base, Philippines. He relished this command at the height of the Vietnam War and was reluctant to leave in July 1968 to become deputy commander of US Strike Command. He retired from that assignment in 1970.

Col Davis with his P-47 as commander of the 332d Fighter Group in Italy.

43

Jacqueline "Jackie" Cochran

- First woman to break the sound barrier.
- First Woman to enter the Bendix Race, 1935.
- Won 14 Harmon Trophies in her lifetime.
- Set altitude record of 33,000 feet, 1939.
- First female trans-Atlantic bomber pilot, 1941.
- Supervised the Women's Air Force Service Pilots (WASP), 1942–44.
- Served as an advisor to the US Air Force, Federal Aviation Administration (FAA), and National Aeronautics Space Administration (NASA) during the '50s and '60s.
- Set world speed record of 1,429 MPH, 1964.

Jackie Cochran was born in Florida into a life of poverty, somewhere between 1905 and 1908. She had almost no formal education and since she was orphaned at birth, she had to depend on foster parents. By the age of 22 she was working at a prestigious beauty salon and was at the top of her profession. With money saved, she developed a line of cosmetics, which would later become her empire, Jacqueline Cochran Cosmetics. Her husband to be, millionaire businessman Floyd Odlum, suggested she learn to fly in order to use her travel and sales time more efficiently. In two days she soloed and 18 days later had her pilots license. Despite her lack of education, she mastered flying in mere weeks. Jackie soon bought her first airplane, a Travelair. She was the first woman to enter the Bendix Race in 1935.

Jackie was hooked on flying and her taste for record setting was strong. She set three speed records, won the Clifford Burke Harmon trophy three times, and set a world altitude record of 33,000 feet—all before 1940. With World War II on the horizon, Jackie talked Eleanor Roosevelt (who, like Jackie, had been friendly with Amelia Earhart) into the necessity of women pilots in the coming war effort. It was probably no small coincidence that Jackie was soon recruiting women pilots to ferry planes for the British Ferry Command, and became the first female trans-Atlantic bomber pilot.

In 1942 Jackie recruited over one thousand WASPs and supervised their training and service until they were disbanded in 1944. She didn't stop there. Jackie went on to be a press correspondent and was present at the surrender of Japanese General Yamashita, was the first US woman to set foot in Japan after the war, went on to China, Russia, Germany, and even the Nuremburg trials. Flying was still her passion, and with the onset of the jet age, there were new planes to fly and records to break! And she did both. Access to jet aircraft was mainly restricted to military personnel, but Jackie had enough connections to get where she wanted to be. With the assistance of her friend Gen Chuck Yeager, Jackie became the first woman to break the sound barrier in an F-86 Sabre Jet, and went on to set a world speed record of 1,429 MPH in 1964. Jacqueline Cochran broke the sound barrier when she was well over 50 years old. After heart problems and a pacemaker stopped her fast-flying activities at the age of 70, Jackie took up soaring. At the time of her death in 1980 she held literally hundreds of speed and altitude records—more than anyone else in the world, male or female.

Gen Leon W. Johnson

- Awarded the Medal of Honor for heroic actions while leading the 44th Bomb Group attacking Ploesti, Romania on 1 August 1943.

Leon Johnson graduated from West Point in June 1926 and went to Fort Crook, Nebraska, as a second lieutenant in the 17th Infantry. After flight training at Brooks and Kelly Air Fields, Texas, he was transferred to the Air Corps with assignment to the 5th Observation Squadron at Mitchell Field. As a lieutenant colonel, he went with Eighth Air Force to England in June 1942 as assistant chief of Staff for Operations. In January 1943, he took command of the 44th Bomb Group, was promoted to colonel and went to Africa on loan to Ninth Air Force.

On 1 August 1943, he led the 44th, called the "Eight Balls," as one of the major elements in the massive B-24 bomber attack on the oil refineries at Ploesti, Romania. The element he led became separated and temporarily lost from the lead elements. Colonel Johnson reestablished contact with the main formation and continued on the mission to discover the assigned target had been attacked and damaged by earlier B-24s. Having lost the element of surprise, upon which the safety and success of such a daring mission depended, Colonel Johnson elected to carry out his planned low-level attack against another target within the Ploesti complex. Despite heavy antiaircraft fire, numerous enemy fighter planes, the eminent danger of exploding delayed-action bombs from the previous element, Colonel Johnson led the 44th forward. Oil fires and intense smoke obscured the target, but Colonel Johnson demonstrated enormous courage and superior flying skill as he led his formation on a bomb run that totally destroyed the important refining plants and installations. He was awarded the Medal of Honor for his heroic leadership.

B-24 Liberators during the low-level bombing of the oil refineries at Ploesti, Romania, 1 August 1943.

Maj Thomas B. "Mickey" McGuire Jr.

- Second highest scoring ace in Air Force history.
- Awarded Medal of Honor for heroic performance on 25 and 26 December 1944.

Thomas "Mickey" McGuire completed high school at Sebring, Florida, and attended Georgia School of Technology. He enlisted at MacDill Field, Florida, on 12 July 1941 and took pilot training at Randolph and Kelly Fields, Texas, getting his wings and commission in February 1942. He then shipped out to the Pacific where he shot down 38 Japanese planes with his P-38 Lightning. In the entire history of the US Air Force, McGuire's 38 aerial victories are second only to Richard Bong's 40 kills.

McGuire went to Australia in March of 1943 and served with the 131st Fighter Squadron, 475th Fighter Group, in the long campaign from Australia to New Guinea, on to Biak Island and the Philippines. He ran up his long score in all types of fighter missions: bombers escort, fighter bomber sweeps, and aerial combat. On Christmas day 1944, McGuire volunteered to lead a squadron of 15 planes escorting heavy bombers attacking Mabalaent Airdrome in the Philippines. As the formation crossed Luzon it was jumped by a large group of enemy fighters. In the ensuing battle, Major McGuire shot down three enemy planes. This was typical of many of his missions. In another action, he scored one Zero and exposed himself to permit a friendly bomber to escape and then shot down three other fighters before the mission was completed.

McGuire lost his life on 7 January 1945 while leading four P-38s over a Japanese-held airstrip on Los Negros Island. A single Japanese fighter jumped the squadron so McGuire led them into a tight Lufbery Circle snaring the Zero inside. The enemy fighter made a sharp dive to get out of the trap, but the P-38s stayed with him from 2,000 feet down to 200 feet. There the formation scattered and the enemy plane maneuvered into position on the tail of one of the Lightnings. The pilot called for help and McGuire responded. However, a tight maneuver with wing tanks still attached caused the fighter to stall at low altitude and crash.

Major McGuire received the Medal of Honor for combat over the Philippine Islands on 25 and 26 December 1944 when he shot down seven enemy fighters in two days, bringing his total to 38. The citation for the medal concludes with a tribute to McGuire's spirit. "With gallant initiative, deep and unselfish concern for the safety of others, and heroic determination to destroy the enemy at all costs, Maj. McGuire set an inspiring example in keeping with the highest traditions of the military service."

Brig Gen Paul W. Tibbets Jr.

- Led a B-17 Squadron on the first B-17 raid on Germany.
- Commanded the B-29 units tasked with modifying the aircraft to carry an atomic bomb and developing tactics and techniques for delivering atomic bombs.
- Dropped the first atomic bomb ever used in combat.

Paul Tibbets was born in Quincy, Illinois, on 23 February 1915. He entered the Army Air Corps on 25 February 1937 at Fort Thomas, Kentucky. After graduating from flying training in February 1938, he was assigned to Fort Benning, Georgia, in the 16th Observation Squadron. After being assigned to the 29th Bomb Group in early 1942, he was selected to command the 340th Bomb Squadron, taking it into combat operations from England in June of 1942. Tibbets flew 25 combat missions in B-17s, including the first American Flying Fortress raid against occupied Europe. Tibbets also flew combat missions in support of the invasion of North Africa before returning to the United States in March of 1943 to participate in the B-29 program. In September 1944 Tibbets was assigned to the Atomic Bomb Project as the Air Force officer in charge of developing an organization capable of employing the atomic bomb in combat operations and mating the development of the bomb to the airplane. On 6 August 1945, Tibbets, commanding the B-29 *Enola Gay*, dropped the first atomic weapon against enemy forces on Hiroshima, Japan. The Japanese surrendered unconditionally on 14 August 1945.

After the war, Tibbets was the technical advisor to the AF commander during the Bikini Bomb Project. From July 1950 to February 1952, Tibbets was the B-47 Project Officer. After serving as 308th Bomb Wing commander, 6th Air Division commander, and deputy director of Operations on the Joint Staff, Tibbets retired as a brigadier general with promotion date effective 4 May 1960.

The *Enola Gay* was one of 15 B-29s modified specifically for the highly secret atomic bomb missions. These airplanes were outfitted with new engines and propellers and faster-acting pneumatic bomb bay doors.

Independent Air Force, Cold War, and Korea

1946–64

X-1 Supersonic Flight (14 October 1947)

The *Glamorous Glennis* piloted by Capt Chuck Yeager, was the first aircraft to break the sound barrier.

On the morning of Tuesday, 14 October 1947, ground crews completed final preparations and hoisted the Bell XS-1 into the bomb bay of a B-29. At 10:00 A.M., the B-29 took off from Muroc's main eight-thousand-foot runway. Aboard the B-29 was 24-year-old Capt Charles E. "Chuck" Yeager. Yeager had made a name for himself as a fantastic stick and rudder test pilot at Wright Field, Ohio, after World War II. During World War II, Yeager had scored 12.5 aerial victories, including five in one day. Yeager had been selected by Col Albert Boyd, chief of the Test Flight Division, because he had demonstrated an uncanny ability to find and understand any airplane's flaws. The XS-1 project engineer was Capt Jack Ridley. Ridley was a test pilot with an advanced degree in aeronautical engineering from the California Institute of Technology.

As the B-29 climbed slowly to altitude, Yeager made his way back to the XS-1's cockpit. Over the weekend, Yeager had broken a couple of ribs falling from his horse near the "Happy Bottom Riding Club." To avoid being grounded by the flight surgeon, he had gone to a civilian doctor off base and only told Ridley to help him figure out how to close the XS-1's door. Ridley provided Yeager with a 10-inch length of broom handle to use as leverage in securing the hatch.

Bob Hoover flew the FP-80 chase plane to a position 10 miles ahead of the B-29 and at an altitude of 40,000 feet. From this position, he would give Yeager an aiming point during his climb. One minute prior to launch, Ridley asked if Yeager was ready. "Hell, yes, let's get it over with." A typical Yeager response. With that, the B-29 went into a shallow dive and at 20,000 feet Ridley pulled the release mechanism and the XS-1 dropped free.

The drop speed was slower than desired, and the XS-1, nicknamed the *Glamorous Glennis* in honor of Yeager's wife, began to stall. Yeager brought the nose down to gain airspeed and then fired all four of his rocket chambers in rapid sequence. Accelerating upward rapidly, Yeager shut down two cylinders and tested the stabilizer control system. Without it he would lose elevator effectiveness and be "in a helluva bind." It was working. He reached Mach 0.92 as he leveled out at 42,000 feet with 50 percent of his propellants left. Igniting the third cylinder pushed the aircraft to Mach 0.98. The machmeter needle fluctuated at this reading momentarily, then passed off the scale. Yeager radioed, "Ridley! Make another note. There's something wrong with this machmeter. Its gone screwy!" Ridley replied, "If it is, we'll fix it. Personally, I think you're seeing things."

Postflight analysis revealed that Yeager had attained a top speed of Mach 1.06, approximately 700 MPH, at 43,000 feet. The sound barrier had been broken. Yeager had felt no wild buffeting or any other indication that he had broken the barrier. Somewhat disappointed he later reported, "it took a damned instrument meter to tell me what I'd done." The flight had lasted only 14 minutes but his "sonic boom" resonated around the world. This major milestone had been accomplished in just over two years in an era without simulator preflights and despite scientists predicting the barrier to be impenetrable. For the newly independent Air Force, created less than a month before, it was a signal for all the greatness to come.

Berlin Airlift (1948–49)

C-47s line up during the airlift operations.

In the spring of 1948, tensions in Europe were high because the United States, Britain, and France were moving toward uniting their three occupation zones in Germany into a single country—West Germany—economically integrated into Western Europe. The Soviet Union instead wanted all four zones united into a single country dominated by the Soviets. The point where the western nations were most vulnerable was West Berlin, an island of western control in the middle of the Soviet zone. In April, the Soviets placed restrictions on surface transportation to the western troop in Berlin for 10 days, but Lt Gen Curtis E. LeMay (US Air Force Europe [USAFE] Commander) provided C-47 transports that supplied the garrisons. In June 1948 the Soviets announced new restrictions intended to cut off all Western ground links to Berlin.

USAFE began supplying the city by air because the western powers were determined to supply the city but did not want a ground confrontation with the Soviets. The airlift began with C-47s but they were soon replaced by four-engine C-54s with three times the cargo capacity. Because of the strategic importance of the airlift, Air Force Chief of Staff, Gen Hoyt S. Vandenberg ordered the formation of a task force headquarters to run the operation under the command of the famous airlifter, Maj Gen William H. Tunner.

The difficulties of the Berlin Airlift were mind-boggling. Air bases spread across Europe participated but regardless of the point of origin, all aircraft were handled by flight controllers at the Berlin traffic center. The coming of winter brought the need to carry tons of coal in addition to food supplies. The Soviets felt the Americans would be unable to supply the needed fuel during the period of bad weather and short days. Despite the poor conditions, US and allied airmen accomplished their missions, often depending on ground controlled approach radar. The radar guided aircraft along a flight path until the pilot could see runway lights. Radar and tight control made it possible to control the steady flow of aircraft, day and night, in good and bad weather.

The USAF's success in the Berlin Airlift thwarted Soviet efforts to intimidate the West in the first "battle" of the Cold War.

The Berlin Airlift frustrated the Soviet blockade by supplying the two million residents of Berlin with food and fuel for almost a year. Instead of discouraging the Western powers, the joint and combined airlift operation strengthened morale in the West. Though the Soviets agreed to lift the blockade in May 1949, the airlift continued until the end of September to ensure a constant supply of fuel and food while ground traffic resumed.

The Berlin Airlift delivered 2.3 million tons of food and fuel. In spite of poor weather and crowded flying conditions, the 15-month operation had an accident rate less than one-half that of the entire Air Force. Gallant efforts by Military Air Transport Service, Air Materiel Command, the Air Weather Service, and USAFE ensured the success of the operation.

Pusan Perimeter Defense (1950)

North Korean T-34 tanks knocked out by US aircraft.

Air Force defensive operations played a key role in the early days of the Korean War against a formidable enemy. Some 30,000 Koreans who fought with Mao Tse-tung in the Chinese communist revolution formed the backbone of the North Korean People's Army or Inmun Gun. After US Secretary of State Dean Acheson declared Korea outside the United State's "defense perimeter" in January 1950, it took only six months for Kim Il Sung to invade South Korea. It took the Inmun Gun only 48 hours to reach Seoul, claiming the city on 28 June 1950. The Inmun Gun stopped for nothing.

General MacArthur, commander in chief of American forces in the Far East, examined the invasion and recommended the United States intervention to President Truman. The North Koreans were trained, supplied, and advised by the Russians; abandoning South Korea to them would have been a dangerous form of appeasement. The United States sent troops into South Korea in July, but they were inexperienced and lacked good equipment. The Inmun Gun moved with minimal opposition toward the South Korean port of Pusan at the southern tip of the peninsula. The only successful opposition to the advance came from the six hundred combat planes of the Far East Air Forces (FEAF).

In early August, North Koreans reached the Naktong River that enclosed the southeastern corner of Korea. Within this 50-mile by 80-mile area was the Eighth Army, commanded by Lt Gen Walton Walker. The Eighth Army faced the gravest challenge of any American ground forces in history. The North Koreans outnumbered the United Nations forces, who had very few reserve units and were spread dangerously thin along the perimeter. However, Eighth Army did have some important advantages, most notably the FEAF. Because of American airpower the North Koreans could only move and fight at night.

The FEAF had severely damaged the Inmun Gun armor to the point of ineffectiveness. Furthermore, no American Army had ever received more close air support. During August,

Korea showed airpower's flexibility when centrally controlled to meet situational needs.

the FEAF flew 7,397 close-air-support missions averaging 238 sorties daily. The flexibility of airpower also brought a decided advantage. When enemy troops achieved penetrations in the perimeter against which no ground strength could be brought to bear, General Walker requested Maj Gen Earl E. Partridge, Fifth Air Force commander, conduct an air attack. On one occasion, massive enemy troop concentrations were targeted by B-29s at MacArthur's request. Ninety-eight B-29's dropped 3,084 five-hundred-pound bombs on Waegwan, destroying the enemy concentration and preventing any major attacks for weeks.

On 31 August 1950, the communists decided to launch an all-out, human wave assault against the Pusan perimeter. During the day, Fifth Air Force fighter-bombers provided 167 close-air-support sorties. The communist attack was repelled prompting Maj Gen William Kean, 25th Infantry Division commander, to remark, "The close air support strikes rendered by the 5th AF saved this division as they have many times before." The Inchon landing finally repelled the North Koreans, but the air operations in the defense of the Pusan Perimeter saved the UN cause.

Corona (1960)

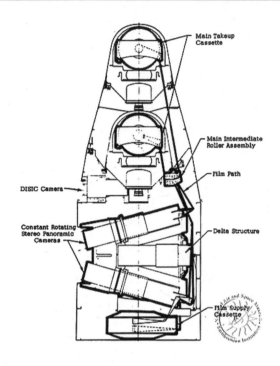

Schematic drawing of a later Corona satellite equipped with two takeup reels and two return capsules.

the photographs could only distinguish objects larger than 40 feet across, this was still an amazing photographic achievement because the camera was more than one hundred miles above the earth and moving more than 17,000 MPH. Since there was no way to electronically transmit the images back to earth, the Corona film had to be physically brought back to the photo analysts.

Once the Corona camera had taken all its pictures, the film was wound onto a reel in the nose cone return capsule. (Later Corona satellites carried twice as much film and two capsules.) The capsule was dropped back to earth at a prearranged time and place. When it had fallen to an altitude of approximately 60,000 feet, it deployed a parachute. Using a radio beacon on the capsule and visually spotting the parachute, aircraft of the 6593d Test Squadron would snag the parachute and reel in the capsule. The film was then analyzed on the ground. Newer systems that have replaced Corona transmit images electronically in real time.

In 1958, a few months after Sputnik, the Soviets demonstrated that they would soon be able to launch nuclear-armed intercontinental ballistic missiles (ICBM) at the United States. President Eisenhower authorized Corona, a top priority reconnaissance satellite program managed jointly by the Air Force and the Central Intelligence Agency (CIA). The Corona program faced three major hurdles: getting the satellite over the Soviet Union, taking useful pictures, and getting the images back to US intelligence analysts. In all of these areas, Corona was at the technological limit of US capabilities.

It took two years and 12 failed launches before Corona finally managed to get a satellite into orbit in August 1960. The next mission (Corona 14) carried the first Corona camera into space and took the first satellite pictures of the Soviet Union. Its photos covered more of the Soviet Union than all the previous U-2 flights combined. Although

Corona was the first Air Force satellite program that accomplished a critical military mission.

Corona images were invaluable for piercing the veil of secrecy surrounding the Soviet military and improved technology increased the quality and quantity of Corona images until the program was superseded in 1972. More than one hundred Corona missions returned over eight hundred thousand photographs to earth. Since Corona, our use of space has increased until it has become vital to virtually all Air Force operations.

Strategic Air Command (1946–92)

Throughout most of SAC's history, the B-52 was the backbone of its manned bomber fleet. The B-52 continues in service today and has been the longest-serving combat aircraft type in USAF history.

Perhaps the greatest achievement of the US military during the Cold War was its success in deterring a Soviet nuclear attack on the United States. The centerpiece of US nuclear deterrence throughout the Cold War was the US Air Force's Strategic Air Command (SAC). Throughout its 46 year life, SAC was the most powerful military force on earth.

SAC was created in 1946, shortly after the end of World War II, to manage what were then the world's only nuclear weapons and nuclear-capable delivery systems (modified B-29s). Though Gen George C. Kenney was the first SAC commander, it was SAC's second commander, Gen Curtis E. LeMay, who truly shaped the character of the command. LeMay commanded SAC from 1948–57—some of the tensest years of the Cold War. LeMay felt that SAC had to be prepared to destroy the Soviet Union at a moment's notice, and he therefore demanded an unprecedented level of combat readiness in peacetime. In 1948 some felt that LeMay was over-reacting but in the early 1950s the Soviets developed nuclear weapons, intercontinental ballistic missiles, and Soviet pilots battled US pilots over Korea.

These events made it clear that the Soviet Union was ready and willing to confront the United States militarily, and made nuclear deterrence an overwhelming national security need.

For more than 40 years, SAC remained continuously on alert with nuclear armed bombers in the air, ready to immediately destroy any adversary. As the United States developed nuclear armed intermediate range missiles, intercontinental ballistic missiles, and cruise missiles, these all enhanced SAC's awesome capabilities. SAC's technological capabilities extended beyond nuclear weapons to surveillance and reconnaissance where it controlled the U-2 and SR-71 and was also an early innovator in using unmanned aerial vehicles for reconnaissance and surveillance.

While SAC is best known for its invaluable service preparing for, and thus deterring nuclear war, it also achieved impressive results in conventional combat. During Vietnam and the Gulf War, SAC bombers delivered devastating conventional strikes against strategic targets and enemy ground forces.

With the collapse of the Soviet Union and the end of the Cold War, SAC's initial mission was complete and US nuclear forces could lower their alert status. In 1992, the US Air Force deactivated SAC as part of Chief of Staff General McPeak's efforts to reorganize the force to meet the challenges of the post-Cold War era. SAC's nuclear deterrent mission went to US Strategic Command, and the personnel and equipment were consolidated with Tactical Air Command (TAC) to form the new Air Combat Command.

The Cuban Missile Crisis (1962)

This soviet freighter photographed in a Cuban port has missile transporters on its deck. The shadow of the RF-101 that took this picture is visible in the lower right of the photo.

In the early 1960s, the nuclear forces of the Union of Soviet Socialist Republics (USSR) were vastly weaker than those of the United States, and the US was trying to overthrow Fidel Castro's Communist government. Cuba and the USSR felt that they could defend Cuba and increase the nuclear threat they posed to the United States by stationing Soviet nuclear armed bombers, medium range missiles, and intermediate range missiles in Cuba. These weapons could hit US cities in a matter of minutes.

In August 1962, the Soviets started building missile sites in Cuba, and in October photos from US Air Force U-2 reconnaissance aircraft revealed what the Cubans and Soviets were up to. President John F. Kennedy called on the joint chiefs of staff (JCS) and his other advisors to consider how the United States should respond. Everyone agreed that the Soviet missiles had to be removed from Cuba but there were several options for forcing the Cubans and Soviets to remove them. One option, favored by the JCS, was to destroy the sites with conventional air strikes followed by a US invasion of the island. President Kennedy felt that this made the risk of nuclear war with the Soviets too likely

and was not absolutely necessary because the missile sites were not yet operational. Instead Kennedy made a televised announcement on 22 October demanding that the Soviets remove the missiles, and announcing that the United States would enforce a blockade on Cuba to prevent the delivery of offensive weapons to the island.

While the president and the national command authorities were weighing their options, US military forces prepared for imminent war. For the Air Force this meant adding RF-101 and RB-66 low-altitude reconnaissance flights to the daily U-2 missions, moving large numbers of fighters to south Florida bases for attacks on Cuba and for air defense of the United States, and airlifting ground forces into Florida. The most dramatic measures involved putting the entire Strategic Air Command (which commanded virtually all US strategic nuclear forces) on full wartime alert. The Soviets responded belligerently and the world came closer to nuclear war than at any other time during the Cold War. Then the Soviets backed down and offered to remove the missiles if the United States agreed not to invade Cuba.

For some cold warriors it was hard to accept the indefinite presence of Communist Cuba just off the southern coast of the United States but President Kennedy accepted the Soviet offer. As the missiles were removed, US forces went back to their normal peacetime bases and activities. Backing down to US demands was a major humiliation for the Soviet Union and forcing them to do so without nuclear war was one of the greatest accomplishments of the Kennedy administration. Kennedy worked from a position of strength because of the power and high readiness of the USAF Strategic Air Command. Spotting the missile sites before they became operational gave Kennedy the option of a blockade and was one of the greatest achievements of USAF reconnaissance.

Brig Gen Charles E. "Chuck" Yeager

- First person to break the sound barrier, 14 October 1947. Piloted the Bell X-1 to Mach 1.05 at 42,000 feet.
- Fighter pilot ace in World War II with 13 and one-half aerial victories to his credit.
- Shot down five enemy aircraft in a single day—12 October 1944.

Chuck Yeager grew up in Hamlin, West Virginia. He enlisted in the Army Air Force in 1941 at the age of 18. In 1943 he became a flight officer (a noncommissioned officer who was allowed to fly) and went to England to fly fighter planes over France and Germany. In his first eight missions, at the age of 20, Yeager shot down two Luftwaffe aircraft. On his ninth mission, his aircraft was hit and he suffered flak wounds forcing him to bail out over France. The French underground smuggled him into Spain, where he found his way back to England and returned to combat in time for the Normandy Invasion.

On 12 October 1944, Yeager engaged five German fighters and shot each of them down. Later that year, flying a P-51 Mustang, he shot down an ME-262 German jet fighter and four FW-190s. By war's end, he had tallied 13 and one-half kills.

After the war, he was trained as a test pilot at Wright Field, Ohio, where he impressed his instructors with his ability to perform stunt flying. Despite his abilities, people were amazed when he was selected to go to Muroc Field, California, for the XS-1 project at the age of 24. On 14 October 1947, despite two broken ribs sustained in a horseback riding accident two days earlier, Yeager climbed into the X-1 *Glamorous Glennis* in the belly of the mother ship B-29. At 26,000 feet the B-29 went into a shal-

low dive, then released the X-1 like a bomb. Yeager ignited the four rocket chambers and began a 45-degree climb. At approximately 10:33 A.M. at 42,000 feet, Yeager made the famous radio call, "Say, Ridley (X-1 Flight Engineer). . . make another note, will ya? There's something wrong with this machmeter . . . it's gone screwy." Yeager had broken the sound barrier achieving Mach 1.05. He remained supersonic for about 20 seconds before cutting off the engine, coasting up to 45,000 feet and then gliding to a landing at Rogers Dry Lake.

Yeager quit testing rocket planes in 1954 and went to Okinawa where he tested a Soviet MiG-15. His assessment confirmed the outstanding characteristics of the MiG-15 but upheld the notion that the F-86 was a comparable aircraft with many advantages. In 1957, Yeager commanded an F-100 squadron at George AFB, California. In 1966, he assumed command of the 405th Fighter Wing at Clark AFB, Philippines. While there he flew 127 missions in South Vietnam. General Yeager's final assignment was director of the Air Force Inspection and Safety Center.

General Yeager won the Mackay and Collier Trophies in 1948 and the Harmon International Trophy in 1954. He retired from active duty on 1 March 1975.

Gen Hoyt S. Vandenberg

- Went from captain to lieutenant general in less than four years.
- During the North Africa campaign of World War II, Vandenberg earned the Silver Star. For his role in planning the Normandy invasion he received an oak leaf cluster for his Distinguished Service Medal.
- Served as director of Central Intelligence from June 1945 to April 1947.
- Served as vice chief under General Spaatz and then succeeded him as Chief of Staff in 1948.
- Served as Chief of Staff during the Berlin Airlift, the deployment of the Hydrogen bomb and the Korean War.

Hoyt S. Vandenberg graduated from the US Military Academy in 1923, and was commissioned a second lieutenant in the Air Service. He graduated from the Air Service Flying School at Brooks Field, Texas, and from the Air Service Advanced Flying School at Kelly Field, Texas, in 1924.

His first assignment was with the 3d Attack Group at Kelly Field, where he commanded the 90th Attack Squadron and developed a reputation as an outstanding pilot and commander. From mid-1934 to 1936 he attended the Air Corps Tactical School at Maxwell Field, Alabama, and the Army Command and General Staff School at Fort Leavenworth, Kansas. With this schooling behind him he became an instructor at the Air Corps Tactical School before attending the Army War College.

After graduating from the War College in 1939, Vandenberg put his training to use in the Air Staff until late 1942 when he was assigned to the United Kingdom. In North Africa he was appointed Chief of Staff of the Twelfth Air Force, which he helped organize. He returned to Washington in 1943 as deputy chief of Air Staff but soon left for the Soviet Union where he helped negotiate the use of bases in the Soviet Union for bombing targets in Eastern Germany out of range of bombers flying from Britain, Italy, and North Africa. These frustrating negotiations, and his participation in the conferences Roosevelt and Churchill held with Chiang Kai-shek in Cairo and with Stalin in Tehran, got Vandenberg involved in wartime diplomacy at the highest levels and helped prepare him for Cold War confrontations with the Soviets.

After a brief return to the Air Staff he went back to Britain to serve as deputy Air Commander in Chief of the Allied Expeditionary Forces and commander of its American Air Component for the invasion of France. Two months after D day Vandenberg assumed command of the Ninth Air Force—the largest tactical air force ever fielded—and commanded it throughout the liberation of France and conquest of Germany. He was promoted to lieutenant general in March 1945, less than four years after he made major.

After the war he served as assistant chief of Air Staff at AAF headquarters for six months before Gen Dwight D. Eisenhower named him director of Intelligence for the War Department. He was so effective in reorganizing Army intelligence that President Harry Truman named him director of the Central Intelligence Group (forerunner of the Central Intelligence Agency).

He returned to the Air Force in 1947 and, as deputy commander and chief of Air Staff, he worked out all the details of Air Force independence from the Army. When the Air Force achieved the independence Vandenberg and so many other airmen had sought for so long, he became the first vice Chief of Staff of the Air Force and was promoted to the rank of general. On 30 April 1948 Vandenberg became Chief of Staff of the Air Force, succeeding Gen Carl A. Spaatz.

During his tour as Chief of Staff he led the Air Force through the early conflicts of the Cold War including the Berlin blockade and the Korean War and struck a balance between conventional tactical forces and strategic nuclear forces. Vandenberg retired from active duty 30 June 1953 and died of cancer the following year.

Col Dean Hess

- Was an ordained minister before joining the Army.
- Flew 63 combat missions in Europe during World War II.
- Flew 250 combat missions in Korea.
- Trained and led Korean pilots.
- Established an orphanage in Korea for the vast number of children orphaned by the war.

On 7 December 1941 Dean Hess was an ordained minister of the Christian Church, living in Cleveland, Ohio. Realizing he could not expect his parishioners to bear arms for the United States if he was not willing to do so, he enlisted in the Aviation Cadet Program and became a pilot. Sent to France in 1944, Hess flew P-47s on 63 combat missions.

Following World War II, Hess returned to the pulpit and to graduate school, but in July 1948 he was recalled to active duty. He was stationed in Japan in June 1950 when the Korean Conflict began and immediately was sent to Korea as commanding officer of a detachment of USAF personnel training South Korean pilots to fly F-51 type fighters. Despite numerous obstacles, Hess was able to expand this training into operational flights, which he often led. By June 1951 when he left Korea, he had flown 250 combat missions. During this period, Hess started an unofficial program for giving food and shelter to the real victims of the war, children who had lost their parents and their homes.

So many were given shelter that Hess's airfield could no longer accommodate them and they were taken to a central orphanage in Seoul, South Korea. When the communists began to overrun the city, Hess persuaded the USAF to lend assistance. In the midst of grave defensive holding actions by United Nations (UN) troops, 15 C-54s were sent to Seoul, under "Operation Kiddy Car," and evacuated hundreds of children to safety. The children were flown to Cheju Island, off the southern coast of Korea, where Hess had established an orphanage. With contributions from UN soldiers, the orphanage was gradually able to accept more and more children. By the time Hess was transferred from Korea, his orphanage had taken in more than 1,054 Korean children who probably would have died without this assistance.

In 1957, Hess published his story in a book entitled *Battle Hymn*. The story was made into a motion picture starring Rock Hudson as Dean Hess. Hess's royalties from both the book and the movie were used to construct a new orphanage near Seoul. Hess retired from the USAF in 1969.

Col James Jabara

The world's first jet-versus-jet ace was James Jabara of the US Air Force who scored his initial victory on 3 April 1951 and his fifth and sixth victories on 20 May 1951. He was then ordered back to the United States for special duty. However, at his own request, he returned to Korea in January 1953. By June he had shot down nine additional MiG-15s, giving him a total of 15 air-to-air jet victories during the Korean War. Jabara was also credited with one and one-half victories over Europe during World War II. (The German Luftwaffe had 22 jet pilot aces during World War II but all claims were Allied prop-driven aircraft.) On 17 November 1966, Jabara, a colonel, was killed in an automobile accident while traveling to a new assignment.

Capt Joseph C. "Joe" McConnell Jr.

The leading jet ace of the Korean War was Capt Joseph McConnell Jr., who scored his first victory on 14 January 1953. In a little more than a month, he gained his fifth MiG-15 victory, thereby becoming an ace. On the day McConnell shot down his eighth MiG, his F-86 was hit by enemy aircraft fire and he was forced to bail out over enemy-controlled waters of the Yellow Sea west of Korea. After only two minutes in the freezing water, he was rescued by helicopter. The following day he was back in combat and shot down his ninth MiG. By the end of April 1953, he had scored his 10th victory to become a "double ace." On the morning of 18 May 1953, McConnell shot down two MiGs in a furious air battle and became a "triple ace." On another mission that afternoon, he shot down his 16th and last MiG-15. A little more than a year later, 25 August 1954, while testing an F-84H at Edwards AFB, California, Capt McConnell crashed to his death.

F-86 Sabres take off heading for "MiG Alley" during the Korean War.

Maj Gen Frederick C. "Boots" Blesse

- Credited with 10 aerial victories during the Korean War, he is America's sixth leading jet ace.
- 1954 and 1955 winner of the Air Training Command (ATC) Fighter Gunnery Championship.
- In 1955, he won all six individual trophies at the Gunnery Championship—a feat that has never been equaled.

Fred Blesse was born on 22 August 1921 in Colon, Panama Canal Zone. He graduated from the United States Military Academy at West Point in 1945. For the next three years, Blesse flew many different aircraft including P-40, P-51, P-47, and F-80. During the Korean War, he flew two volunteer combat tours. He completed 67 missions in F-51 aircraft, 35 missions in F-80 aircraft, and 121 missions in F-86 aircraft. During his second combat tour, he was credited with destroying nine MiG-15s and one LA-9. He was the Air Force's leading jet ace when he returned to the United States in October of 1952.

In December of 1952, Blesse was assigned to Nellis AFB, Nevada, as a jet fighter gunnery instructor, squadron operations officer, and squadron commander. In the 1955 Air Force Worldwide Fighter Gunnery Championship, Blesse flew an F-86F and claimed all six trophies offered for individual performance—a feat that has never been equaled. While at Nellis, Blesse wrote a fighter tactics book, *No Guts, No Glory,* that was used worldwide to teach fighter operations. As recently as 1973, three thousand copies were sent to tactical units in the field.

In April 1967, Blesse was assigned as director of Operations for the 366th Tactical Fighter Wing at Da Nang AB, Republic of Vietnam. During this one-year tour, he flew 108 combat missions over North Vietnam and another 46 in Laos and South Vietnam. In 1968, he was assigned to Nellis AFB as the director of Operations for the first F-111 wing and in June of 1969 became its commander.

In August 1974, General Blesse was appointed deputy Inspector General of the United States Air Force. He accumulated more than sixty-five hundred flying hours and 650 hours of combat flying. He is the sixth ranking American jet ace.

Mikoyan and Gurevich's (MiG) 15 was the premier communist fighter during the Korean Conflict. Above 27,000 feet it was faster and more maneuverable than the F-86.

61

Brig Gen Robinson "Robbie" Risner

1965, Risner was shot down over North Vietnam and rescued. In September of 1965, he was shot down again over North Vietnam and captured. While prisoner in Hanoi, North Vietnam, Risner served first as the Senior Ranking Officer and later as vice commander of the 4th Allied Prisoner-of-War Wing. He was repatriated in February 1973. Risner was promoted to brigadier general effective 8 May 1974. His military decorations include the Air Force Cross with one oak leaf cluster, Silver Star with one oak leaf cluster, Distinguished Flying Cross with two oak leaf clusters, Bronze Star with V device and one oak leaf cluster, Air Medal with seven oak leaf clusters, and the Purple Heart with three oak leaf clusters. General Risner retired on 1 August 1976.

Robinson Risner was born in Mammoth Spring, Arkansas, on 16 January 1925. He enlisted in the Army Air Corps in April of 1943 and was awarded pilot wings and his commission in May of 1944. After training in P-38s and P-39s and serving in Panama, he was relieved of active duty in 1946 and became an F-51 pilot in the Oklahoma Air National Guard. Called back to service in 1952, he was assigned to Kimpo, Korea, with the 4th Fighter Wing. He flew one hundred combat missions in F-86s and is credited with eight aerial victories to become the 20th jet ace during the Korean War. While assigned as a squadron commander at George AFB, California, he was selected to fly the Charles Lindbergh Commemoration Flight from New York to Paris. Flying the F-100, he set a transatlantic speed record, covering the distance in six hours and 38 minutes. In August 1964 Risner was assigned commander of the 67th Tactical Fighter Squadron at Kadena AB, South Korea, flying the F-105 Thunderchief. While TDY to Thailand in April of

The F-86 Sabre combined with the pilots' skills gave Americans a 792 to 78 shoot down advantage over the MiG-15 during the Korean Conflict from December 1950 through the summer of 1953.

Gen Nathan F. Twining

- In 1957 he became the first Air Force Chief of Staff to serve as chairman of the Joint Chiefs of Staff.
- Began his active service in June 1916 with Company H of the Third Oregon Infantry (National Guard) and served as a corporal on Mexican border duty.
- In 1943, he was placed in tactical control of all Army, Navy, Marine, and Allied Air Forces in the South Pacific, one of the first Joint Air Commands in US history.
- On 2 August 1945 he was appointed commander of Twentieth Air Force in the Pacific; a few days later, his command dropped the first atomic bomb, at Hiroshima.

Nathan F. Twining came from a rich military background; his forebears had served in the American Army and Navy since the French and Indian Wars. Twining enlisted in World War I but soon received an appointment to West Point. Because the program was shortened to produce more officers for combat, he spent only two years at the academy. After graduating in 1919 and serving in the infantry for three years, he transferred to the Air Service. Over the next 15 years he flew fighter aircraft in Texas, Louisiana, and Hawaii, while also attending the Air Corps Tactical School and the Command and General Staff College.

When war broke out in Europe he was assigned to the operations division on the Air Staff; then in 1942 he was sent to the South Pacific where he became chief of staff of the Allied air forces in that area. In January 1943, he assumed command of the Thirteenth Air Force, and that same November he traveled across the world to take over the Fifteenth Air Force from Jimmy Doolittle.

When Germany surrendered, Arnold sent Twining back to the Pacific to command the B-29s of Twentieth Air Force in the last push against Japan, but he was there only a short time when the atomic strikes ended the war. He returned to the States where he was named commander of Air Materiel Command, and in 1947 he took over Alaskan Air Command. After three years there he was set to retire as a lieutenant general, but when Muir S. Fairchild, the vice chief of Staff, died unexpectedly of a heart attack, Twining was elevated to full general and named his successor.

When Vandenberg retired in mid-1953, Twining was selected as chief; during his tenure, massive retaliation based on airpower became the national strategy. In 1957, President Eisenhower appointed Twining chairman of the Joint Chiefs, the first airman to hold that position.

Lt Gen William H. Tunner

- Greatest Airlift commander in USAF History.
- Commanded the "Hump" Operations in World War II.
- Headed the task force that ran the Berlin Airlift.
- Commanded airlift operations during the Korean War.

Gen William H. Tunner was born in 1906 and graduated from West Point in 1928. He completed advanced flying school in 1929 and served in various Army Air Corps units throughout the 1930s, but he made his reputation as an airlift genius during World War II.

General Tunner's first major assignment of the war was as commander of the Ferry Division of Air Transport Command. At that time the Ferry Division was ferrying more than 10,000 aircraft a month to users in the continental United States and overseas. From there he took command of the "Hump" airlift: the toughest airlift operation of the war.

When the Japanese cut the overland route between Allied forces in China and Burma, the forces in China had to be supplied by air from India over the Himalayas. This route was called the "Hump" and the entire operation defied common sense. The austere "base" areas in northern India were about as far from the United States as any place on earth; the weather over the Himalayas was treacherous; the navigational equipment was primitive; the pilots were

inadequately trained; and the unpressurized, piston-engine aircraft could barely get their loads over the mountains. But General Tunner instituted maintenance, organization, and crew qualification programs that drastically reduced the accident rate while moving more cargo than any previous airlift.

When the Soviets closed the land routes to Berlin in 1948, the US Air Force called on General Tunner to run the Berlin Airlift. Once again Tunner streamlined and standardized maintenance and operational procedures driving up the cargo tonnage delivered while maintaining an extraordinarily low accident rate. In 1950, General Tunner was called upon to direct airlift operations in Korea where he made Combat Cargo Command into a remarkably flexible and efficient theater airlift organization and made history by parachuting in a bridge that enabled the US Marines to get over Funchilin Pass.

After the Korean War General Tunner served as commander of US Air Forces Europe and then as commander Military Air Transport Service (a predecessor of Air Mobility Command) before retiring in 1960.

Gen Curtis E. LeMay

- Commissioned through the Reserve Officer Training Corps (ROTC) at Ohio State University in 1928.
- Began World War II as a group commander in Eighth Air Force, but within 18 months had gone from lieutenant colonel to major general and an air division commander.
- Commanded the B-29 bombing campaign against the Japanese home islands.
- In 1948, he commanded US Air Forces in Europe (USAFE) and initiated the Berlin Airlift.
- Commander of SAC from 1948–57.
- After serving as vice Chief of Staff, he became Chief of Staff in 1961.

Curtis E. LeMay, one of the icons of American military history, rivals Mitchell in importance and controversy. Early in his career he became known as one of the best navigators and pilots in the Air Corps. War brought rapid promotion and increased responsibility. He had earned a reputation as an unusually innovative tactician and problem solver, so when Hap Arnold had difficulty making the new B-29 combat effective, he chose LeMay to take over B-29 operations in China. His ability led Arnold to name him commander of the XXI Bomber Command conducting the main air effort against Japan from the Marianas. Always a tactical innovator, LeMay took the risky and controversial step of abandoning the long-held American doctrine of high altitude, daylight, precision bombing, and instead stripped his B-29s of guns, loaded them with incendiaries, and sent them against Japanese cities at night and at low level. The new strategy was remarkably successful; Japan was devastated, and then with the dropping of the atomic bombs in August 1945 brought the Pacific war to an end.

In 1948, LeMay was sent to Germany to command the air forces in Europe arrayed against the Soviets. In this position he was responsible for getting the Berlin Airlift started in 1948. This crisis precipitated a major reshuffling in Washington. A war with the Soviets appeared increasingly possible, and the Strategic Air Command, which would bear the brunt of such a war, was seen as deficient. As a result, Hoyt S. Vandenberg relieved George C. Kenney from command at SAC and named LeMay his successor. The building of SAC into an effective and efficient warfighting arm was LeMay's greatest accomplishment. To test his command's state of readiness he ordered a "bombing raid" on Dayton, Ohio, but not a single SAC aircraft carried out the mission as planned. He then set about the difficult but essential task of retraining SAC. Using the authority delegated him by Vandenberg, LeMay built new bases, facilities, and training programs; began a "spot promotion" system for rewarding his best aircrews; and, through his legendary use of iron discipline, soon transformed his command into one of the most effective military units in the world.

In 1957, LeMay was named vice Chief of Staff, and when Thomas White retired in 1961, he was elevated to the position of chief. LeMay had strong feelings regarding American involvement in Vietnam, arguing against the gradual response advocated by the administration. He retired in 1965.

Gen Bernard A. "Bennie" Schriever

- The "Father" of the Ballistic Missile Program.
- In August 1954 he was appointed commander of the Air Force Western Development Division, ARDC, Inglewood, California.
- Directed the nation's highest priority project in the 1950s—the development of a ballistic missile program.
- He was responsible for technical phases of the Atlas, Titan, Thor, and Minuteman ballistic missiles.
- Assumed command of Air Force Systems Command in April of 1961.

Bennie Schriever was born in Bremen, Germany, on 14 September 1910. His parents migrated to the United States in 1917 and he became a naturalized citizen in 1923. He graduated from Texas A&M in 1931. Schriever entered flight training at Randolph Field, Texas, in June of 1932 and graduated in 1933.

In 1939 Schriever was assigned as a test pilot at Wright Field, Ohio. After receiving advanced schooling in engineering at Stanford University, he was assigned to the 19th Bomb Group in the Southwest Pacific in July 1942. While in the Pacific theater he participated in the Bismarck Archipelago, Leyte, Papua, North Solomon, South Philippine, and Ryukyu campaigns.

From 1946 to 1954 Schriever served in several positions at the headquarters concerning material and development planning. In August 1954, he was appointed commander of the Air Force Western Development Division, ARDC, Inglewood, California. In this capacity, he directed the nation's highest priority project—the development of a ballistic missile program and the initial space programs. He was responsible for technical phases of the Atlas, Titan, Thor, and Minuteman ballistic missiles, as well as providing the launching sites and equipment,

tracking facilities, and ground support necessary for missile operation.

Schriever assumed command of Air Force Systems Command in April of 1961 and was promoted to general in July 1961. He retired from active duty on 1 September 1966.

The first Titan intercontinental ballsitic missile launch on 6 February 1959.

Lt Col Edwin "Buzz" Aldrin

Edwin "Buzz" Aldrin was an American astronaut born in Montclair, New Jersey. He was nicknamed Buzz by his sister. He trained at West Point, flew combat missions in Korea, and later flew in Germany with the 36th Tactical Wing. He logged over twenty-nine hundred hours of flying time. Aldrin became an astronaut in 1963. In 1966 he set a world record by walking in space for five hours and 37 minutes during the Gemini 12 mission, the last Gemini mission. He was the second man to set foot on the moon in the Apollo 11 mission in 1969.

Lt Col Virgil I. "Gus" Grissom

Gus Grissom was an American astronaut who was born on 3 April 1926 in Indiana. Before he became an astronaut, Grissom served as a fighter pilot in the Korean War. He logged over forty-six hundred hours of flight time. Grissom became an astronaut in 1959. He was one of the original seven Mercury astronauts. He spent over five hours in space on two spaceflights. In 1961, Grissom flew aboard Mercury 4. He became the second American to make a suborbital flight into space during this 16-minute flight. In 1965 Grissom piloted the first maneuverable spacecraft aboard Gemini 3. Grissom was one of three astronauts to be killed in the Apollo I fire in 1967. The three astronauts were training for the first manned Apollo flight when an electrical fire broke out in the cockpit.

Maj Donald K. "Deke" Slayton

Donald "Deke" Slayton, an American astronaut, was born on 1 March 1924 in Wisconsin. Before he became an astronaut, Slayton was a fighter pilot in World War II. He logged over nine thousand hours of flight time.

Slayton became an astronaut in 1959. He was one of the original seven Mercury astronauts. He spent over 216 hours in space on one spaceflight. In 1975 Slayton flew aboard the Apollo-Soyuz Test Project. The Apollo spacecraft linked up with a Soviet spacecraft. The crews of each country spent nine days performing experiments together. Slayton left NASA in 1981.

Chief Master Sergeant of the Air Force Paul W. Airey

1944–May 1945), following capture after he was forced to bail out of his flak-damaged aircraft.

During the Korean Conflict, Chief Airey was awarded the Legion of Merit while assigned at Naha Air Base, Okinawa. The award, an uncommon decoration for an enlisted man, was earned for having conceived a means of constructing equipment from salvaged parts that improved corrosion control of sensitive radio and radar components.

CMSgt Paul W. Airey was selected from among 21 major air command nominees to become the first Chief Master Sergeant of the United States Air Force. He was formally installed on 3 April 1967, by Gen John P. McConnell, Air Force Chief of Staff. His previous assignment was with the Air Defense Command's 4756th Civil Engineering Squadron at Tyndall AFB, Florida, where he served as the unit's first sergeant.

Chief Airey spent much of his 24-year career as a first sergeant. During World War II, however, he served as an aerial gunner on B-24 bombers, and is credited with 28 combat missions in Europe. Chief Airey was a prisoner of war in Germany (19 July

Like the B-24 in this photograph, Airey's B-24 was brought down by German flak.

THE VIETNAM WAR

1965–75

Operation Rolling Thunder (1965–68)

The F-105 Thud delivered 75 percent of bomb tonnage during Operation Rolling Thunder.

The Rolling Thunder air campaign lasted from February 1965 to March 1968. To avoid escalating military actions until after the 1964 elections, President Lyndon B. Johnson did not send significant US ground forces to Vietnam until July 1965. The US Military Assistance Advisory Group, Vietnam, had been operating in Southeast Asia since 1955 but had not prevented the Vietnamese communists from making steady progress against South Vietnamese forces. By January 1965, Johnson was under mounting pressure from his closest national security advisors to "do something" in response to the increasing communist attacks on United States and South Vietnamese personnel and installations.

Secretary of Defense Robert S. McNamara spelled out the US strategic objective in a memorandum for the president. In McNamara's words, "Our objective is to create conditions for a favorable settlement by demonstrating to the [communists] that the odds are against their winning." The administration believed that once the communist North Vietnamese realized they probably would not win, they would stop attacking the South. Johnson's advisors considered three options: continue a policy of limited military assistance; launch bold attacks against the North; or gradually increase the use of military force.

Option three, perceived as a middle ground between the hawks and doves, was endorsed by the president. To the greatest extent possible, the administration wanted to achieve its ends through a finely calibrated use of airpower known as "Rolling Thunder." Fearing that the Air Force would be too aggressive, the White House maintained tight control of all phases of the operation. Johnson correctly claimed that the military could not bomb an outhouse in North Vietnam without his permission. In the words of one historian, "the policy-makers did everything but fly the aircraft." But the campaign failed to achieve US policy objectives.

Top strategists within the administration noted the ineffectiveness of the air campaign from the beginning. For example, in April 1965 the chairman of the Joint Chiefs of Staff (JCS) (Gen Earle G. Wheeler) told Secretary McNamara that the February–April bombings had not appreciably reduced "the overall military capability of the [communists]. I think it is fair to state that our strikes to date, while damaging, have not curtailed [their] military capabilities in any major way."

Rolling Thunder's use of precise increments of force succeeded only in preventing the attainment of US strategic objectives. In fact, these halfhearted attempts at using airpower bolstered the enemy's resolve.

Rolling Thunder demonstrated US support for the government of South Vietnam and the United States' desire to contain communist expansion but it failed to bring the North Vietnamese to the negotiating table, failed to impede the flow of logistics support from the North, and failed to weaken the North's resolve. In all, the United States flew 304,000 fighter-bomber sorties and 2,380 B-52 bomber sorties over North Vietnam, lost 922 aircraft, and dropped 634,000 tons of bombs. The failure of Rolling Thunder also contributed to the decisions that sent US ground troops to Vietnam, escalating the war and raising the price of our ultimate failure in Vietnam.

Operation Arc Light (1965–67)

The B-52 Stratofortress proved its worth in Vietnam.

The United States began direct involvement in military operations in Southeast Asia in the spring of 1965. In April, Gen William C. Westmoreland, commander of Military Assistance Command, Vietnam (MACV), urged the use of B-52s against Vietcong base camps in advance of ground forces. These base camps were too spread out to be effectively bombed by the more tactical fighter-bombers. B-52s could deliver an even pattern of bombs over a large area in a short time. The Joint Chiefs approved the idea and by May the details for the strikes had been worked out. The first strike was scheduled for 18 June 1965.

At 1:00 A.M., 30 B-52s took off from Guam and headed to the refueling area off the northwestern tip of Luzon in the Philippines. Most of the bombers carried 51 750-pound bombs. Six of the B-52s were loaded with one- thousand-pound armor-piercing bombs. A tailwind caused the bombers to arrive at the rendezvous early so the lead cell of 10 aircraft began to circle. In the dark, the turn caused two of the Stratofortresses to collide killing eight of the 12 crewmembers. Twenty-seven of the remaining bombers refueled and continued the mission.

At 6:45 A.M. and between 19,000 and 22,000 feet, the B-52s began dropping their bombs on a one-mile-by-two-mile target box. Slightly more than one-half of the thirteen hundred bombs hit the target box. Shortly after the B-52s departed, 32 A-1Es from Bien Hoa strafed the base camps to soften them up as landing zones. Then three reconnaissance teams performed a ground sweep of the target area but found no enemy and little bomb damage. The B-52s returned to Guam, completing the 13-hour mission. The immediate observable results of the bombing were less than spectacular especially placed in the politically charged atmosphere of the time. Journalists questioned why such a powerful strategic weapon was used on such a minuscule enemy. However, the B-52s were one part of a large operation that included tactical aircraft, ground troops, and helicopters. The B-52s added an element to the air war that had been missing. They could strike dug-in targets, saturate large areas, surprise the enemy, reduce enemy safe areas, and encourage the South Vietnamese soldiers to venture back into enemy territory.

The tactic of using B-52s to surprise enemy concentrations and soften them for a follow-up ground sweep saw a steady increase in use after this first mission in 1965. By July 1966, B-52s had flown 4,309 sorties in 471 missions. Westmoreland used subjective valuations of the missions (weight of ordnance dropped, percent of bombs planned and dropped) to press for increases. He viewed the missions as extremely effective in lowering enemy morale, increasing desertions, forcing changes in tactics, and causing disruptions in the enemy's economy.

American ground commanders considered the B-52s as the most effective weapon system used in South Vietnam. Their enthusiasm fell in line with their view that bombers were "flying artillery." The Air Force objected to the B-52's Arc Light role because the strategic assets were not being used effectively. The draw for politicians was clear due to the high enemy attrition rate and an extremely low loss of US life. During 1965 through 1967, the Air Force flew 25 percent of its tactical strike sorties (46,000) and 30 percent of its B-52 sorties (3,300) in supporting ground offensives.

Operation Bolo (January 1967)

Col Robin Olds, an ace in World War II, won four victories in Vietnam.

On 2 January 1967, 12 F-4s of the 8th Tactical Fighter Wing shot down seven MiG-21s over North Vietnam without a loss. The commander of the 8th, Col Robin Olds, led the sweep and got one of the kills. Nearly two years into the Rolling Thunder air campaign against North Vietnam, Operation Bolo was the first time the Air Force managed to destroy more than two enemy aircraft on a single day.

By the end of 1966, President Johnson had not yet permitted air strikes on North Vietnamese airfields, where Soviet and Chinese advisors were staying. Flying from these sanctuaries, MiG pilots attacked F-105s laden with bombs. Although most F-105s survived such attacks, North Vietnamese pilots often succeeded in getting F-105 pilots to jettison their bombs. Meanwhile, MiG pilots carefully avoided combat with F-4 escorts, screens, and sweeps.

Colonel Olds took advantage of the introduction of electronic jamming pods on F-105s to fool MiG pilots and their controllers into thinking that F-4s were actually more vulnerable F-105s. The new pods protected F-105s against surface-to-air missiles by jamming North Vietnamese radar. During a New Year's Day cease-fire, Olds had the pods transferred from F-105 bases to F-4 bases at Ubon, Thailand, and Danang, South Vietnam. On 2 January, 28 F-4s flew north with jamming pods which made them appear to be F-105s and enticed MiG-21s at Phuc Yen into the air.

Olds flew lead for one flight of F-4s, his first mission to Hanoi. The Hanoi Delta was a dangerous place to loiter particularly for the 55 minutes called for by the plan. As Olds pushed his flight deep into North Vietnam airspace, MiGs began to scramble from every base. Col Daniel "Chappie" James Jr., flying lead on another flight spotted a MiG closing on Olds and warned him. Soon there were numerous MiGs all over the sky and a no-holds-barred dogfight ensued. The engagement lasted only nine tense minutes before the MiGs dived through the clouds and landed at Phuc Yen—saved from further losses by President Johnson's rules of engagement.

The Vietnamese had launched 32 enemy fighters and the F-4s had downed seven of them. If the weather had cooperated and all the MiGs been destroyed, mission planning would have most certainly been affected. As it was, Seventh Air Force returned to business as usual shortly after the success of Operation Bolo. Not until the spring did Johnson authorize attacks on North Vietnamese airfields, and not until the fall did he permit Phuc Yen airfield to be struck—causing surviving North Vietnamese MiGs to be based in neighboring China for the remaining months of Rolling Thunder.

In mid-1967, the F-4C kill ratio was an impressive 21-to-1 against the MiG-21. By 1973, even with improved F-4E aircraft, missiles, and gun sights, the ratio had dropped to 1.06-to-1 against the same MiG-21s.

The F-4 Phantoms claimed 107.5 of the 137 aerial victories of the Vietnam War.

Operation Linebacker I (1972)

This F-4 is carrying 2,000 lb laser-guided bombs to strike strategic targets in North Vietnam in 1972.

By the spring of 1972, the United States had withdrawn virtually all of its ground troops from South Vietnam. On 29 March 1972 the North Vietnamese launched a major invasion of South Vietnam expecting to have an easy time against the South Vietnamese army and hoping to influence the November 1972 US presidential election in their favor. Unlike the guerrilla warfare that characterized much of the Vietnam War, the Communist Spring Offensive of 1972 was a conventional invasion by 14 divisions and 26 separate regiments led by hundreds of tanks and supported by ferocious artillery barrages. The United States immediately responded with airpower in an operation which became know as Linebacker I.

Like the US ground presence in Vietnam, the US air presence had been radically scaled back by early 1972. To meet the communist aggression, the US Air Force rapidly deployed forces into the region. Soon the USAF had very powerful forces in the area including over 200 B-52s in theater and almost 400 F-4s operating out of Thailand and South Vietnam.

In some of the most effective counterland operations of the war, US Air Force B-52s chewed-up the massed communist forces, and F-4s used laser-guided bombs in interdiction missions that cut communist supply lines. C-130s dropped supplies to besieged towns and at night, AC-119 and AC-130 gunships provided vital close air support to hard pressed garrisons. The devastating US airstrikes gradually turned the tide of the fighting and enabled the South Vietnamese to retake every important town that the communists had managed to capture.

One of the striking departures from earlier Rolling Thunder campaign was in strategic attack. Improved diplomatic relations with China enabled US aircraft to attack important targets that have previously been off limits out of concern that attacking them might cause the Chinese to intervene on the ground. At the same time, laser-guidance kits made USAF bombs 100 times as accurate as they had been a few years earlier. These two changes enabled fast-moving F-4s to strike key road and rail targets near the Chinese border efficiently and effectively. In North Vietnam, strategic attacks hit more than 400 bridges, cut key mountain roads in over 800 places, demolished petroleum storage facilities, power-generating plants, headquarters, and repair facilities (the North Vietnamese did not have any factories). Combined with mine-laying by US Navy planes in North Vietnamese harbors, these strategic attacks finally started to isolate North Vietnam from outside support.

This North Vietnamese attempt at a conventional offensive (requiring massed forces and enormous quantities of supplies) came at a time when the United States was improving relations with North Vietnam's ally, China. This mistake made the North Vietnamese army an early victim of the counterland and strategic attack operations that make up the USAF's "rapid halt" strategy.

Operation Linebacker II (1972)

The B-52s of Linebacker II conducted the most concentrated bombing campaign of the war.

Not since World War II had bombers been employed in an operation the size of Linebacker II. In December of 1972, the United States position in Southeast Asia was precarious. The North Vietnamese had successfully defeated the South Vietnamese forces and were unwilling to pursue peace negotiations. The American ground presence had been reduced to approximately 26,000. A force this size could not engage the communists in any large battle. President Richard M. Nixon turned to the Air Force for a solution.

The resultant operation, Linebacker II, occurred 18–29 December 1972, standing down for Christmas day. The operation would be fought Air Force style—a concentrated, sustained air attack against the enemy center of power.

The operation would be very risky because the enemy had massed an awesome air defense system around the Hanoi-Haiphong area. The system included 145 MiG fighters, 26 SA-2 surface-to-air missile (SAM) sites, concentrated antiaircraft artillery and a very sophisticated radar network. The mission was to destroy all major target complexes in the Hanoi-Haiphong area using B-52s and F-111s during the night and tactical aircraft during the day.

The flight line at Andersen AFB, Guam, was crowded with 99 B-52Gs and 50 B-52Ds. The mission from Guam would take 12 hours and require in-flight refueling. At U Tapao Royal Thai Airfield, Thailand, there were another 54 B-52Ds available. Missions from Thailand would only take three to four hours and not require refueling. It took nearly two hours for the 87 B-52s from Guam to taxi, take off, and become airborne on the afternoon of 18 December 1972. They were joined in flight by 42 B-52s from Thailand. Targets on the 18th included airfields, vehicle repair centers, rail yards, railroad repair facilities, and the main Hanoi radio station.

On day one of the operation, the North Vietnamese launched two hundred SAMs often in 4-6 weapon volleys. United States losses that day totaled three B-52s and one F-111. SSgt Samuel Turner, tail gunner on Brown 03, shot down a MiG-21, the first in B-52 combat history.

Tactics were revised for day two—altitudes were lowered and time separation of aircraft was increased to four minutes. The changes resulted in no aircraft losses despite the North Vietnamese shooting over 180 SAMs. Combat losses for the entire campaign totaled 15 B-52s, two F-111s, and two F-4 flying MiG combat air patrol (MiGCAP). The B-52s had proven to be an effective force able to meet and defeat the enemy. Within 60 days of the Paris Peace Accords signing, 591 American POWs were released.

In Linebacker II, B-52s had flown 729 of 741 planned sorties and dropped 15,000 tons of bombs. The North Vietnamese fired 1,240 surface-to-air missiles but the B-52s only suffered a 2 percent loss rate. Not long after Linebacker II, the United States formally disengaged from the war in Southeast Asia. Many believe a bombing campaign like Linebacker II should have been conducted in 1965. In 1965, the Hanoi air defense system was severely limited and an operation of this nature would have achieved a military victory and prevented the costly US involvement in Vietnam.

A1C William H. Pitzenbarger

On 11 April 1966, 21-year old A1C William H. Pitzenbarger of Piqua, Ohio, was killed while defending some of his wounded comrades. For his bravery and sacrifice, he was posthumously awarded the Medal of Honor.

"Pits," as he was known to his friends, was nearing his three hundredth combat mission on that fatal day when some men of the US Army's 1st Division were ambushed and pinned down in an area approximately 45 miles east of Saigon. Two HH-43 "Huskie" helicopters of the USAF's 38th Aerospace Rescue and Recovery Squadron were rushed to the scene to lift out the wounded. Pits was a pararescueman (PJ) on one of them. Upon reaching the site of the ambush, Pits was lowered through the trees to the ground where he attended to the wounded before having them lifted to the helicopter by cable. After six wounded men had been flown to an aid station, the two USAF helicopters returned for their second loads. As one of them lowered its litter basket to Pitzenbarger, who had remained on the ground with the 20 infantrymen still alive, it was hit by a burst of enemy small-arms fire. When its engine began to lose power, the pilot realized he had to get the Huskie away from the area as soon as possible. Instead of climbing into the litter basket so he could leave with the helicopter, Pits elected to remain with the Army troops under enemy attack and he gave a "wave-off" to the helicopter which flew away to safety.

Pits continued to treat the wounded and, when the others began running low on ammunition, he gathered ammo clips from the dead and distributed them to those still alive. Then, he joined the others with a rifle to hold off the Vietcong. About 7:30 P.M. that evening, Bill Pitzenbarger was killed by Vietcong snipers. When his body was recovered the next day, one hand still held a rifle and the other a medical kit.

An HH-43 Huskie on rescue alert could be airborne in about one minute. It carried two rescuemen/firefighters and a fire suppression kit.

Lt Col Merlyn H. Dethlefsen

On 10 March 1967, Merlyn H. Dethlefsen was flying an F-105F as No. 3 in a four-plane formation on a mission against the Thai Nguyen Steel Mill, 50 miles north of Hanoi. The task for the F-105s was to knock out surface-to-air (SAM) and antiaircraft gun sites before the bombing forces arrived. When the four F-105s made their low-altitude attack run, the flight leader was shot down and No. 2 was damaged so heavily that he had to head homeward. Although standard tactics called for only one attack pass on such a heavily defended area, or two at the most, Lt Col Dethlefsen decided not to leave the area, but to continue his attacks. However, a MiG appeared and he had to fly through heavy antiaircraft fire to escape from the MiG; in doing so, his F-105 was also hit and seriously damaged. Instead of heading for home, he elected to carry-on and even after the steel mill had been bombed and the bombing force had withdrawn, he, along with his wingman, stayed in the target area looking for SAM sites. After evading a second MiG and then having his F-105 hit once again by flak, he spotted two SAM sites and attacked, destroying them both. Only then did he head his battered F-105 for friendly territory. For his valor in combat above and beyond the call of duty, Dethlefsen was awarded the Medal of Honor, the third USAF member to receive the award during the South East Asian (SEA) conflict.

The F-105 Thunderchief was called the "Thud" by its aircrews.

Capt Lance P. Sijan

An F-4 on a bomb drop during the Vietnam War.

On 9 November 1967, the F-4 flown by 1st Lt Lance P. Sijan of Milwaukee, Wisconsin, was hit by North Vietnamese ground fire and exploded. Although badly wounded, he was able to parachute from his stricken plane. Without food and very little water he managed to avoid capture for 45 days. A serious compound fracture of the left leg made it impossible for him to walk, but he did manage to pull himself backward through the jungle. Even with the broken leg, a skull fracture, and a mangled right hand, he was able to escape shortly after his initial capture. Recaptured, he was taken to Vinh and thrown into a bamboo cell. He was "interrogated" repeatedly, and in spite of his captors technique of twisting his damaged right hand he refused to disclose any information but his name. When not being interrogated he attempted additional escapes, with beatings the only results. Lt Sijan was soon moved to a POW camp at Hanoi. Even in his pitiful condition, he attempted more escapes, all meeting with failure. His physical condition continued to weaken without proper food or medical attention and soon he developed respiratory problems. After many months of ill treatment, his health—but not his spirit—broken, he was dragged one night from his cell while delirious and never seen again.

Lt Sijan was promoted posthumously to captain on 13 June 1968. On 4 March 1976 he was awarded posthumously the Medal of Honor that was presented to his parents by President Gerald R. Ford. Captain Sijan thus became the first graduate of the US Air Force Academy to receive our nation's highest decoration for heroism above and beyond the call of duty.

Gen John D. Ryan

- In April 1946 Gen John D. Ryan assumed duties with the 58th Bombardment Wing and participated in the Bikini Atoll atomic weapons tests.
- In December 1962 General Ryan joined a select group of athletes when he was chosen a member of the Sports Illustrated Silver Anniversary All-American team.
- Appointed vice Chief of Staff of the US Air Force in August 1968, and Chief of Staff of the US Air Force in August 1969.

General Ryan was born in Cherokee, Iowa, 10 December 1915. Following graduation from Cherokee Junior College in 1934, he entered the United States Military Academy from which he graduated in 1938. He next attended flying school at Randolph and Kelly Fields, Texas, and received his pilot wings in 1939.

In February 1944, he was transferred to Italy where he commanded the 2d Bombardment Group and later became operations officer for the 5th Bombardment Wing, Fifteenth Air Force. He returned to the United States in April 1945, and became deputy Air Base Commander, Midland Army Air Field, Texas. In September 1945, he was assigned to the Air Training Command (ATC) at Fort Worth and Randolph Field, Texas, where he remained until April 1946 when he assumed duties with the 58th Bombardment Wing and participated in the Bikini Atoll atomic weapons tests.

Ryan became director of Materiel for the Strategic Air Command in June 1956, and four years later assumed command of SAC's Sixteenth Air Force in Spain. In July 1961 he was named commander of the Second Air Force at Barksdale AFB, Louisiana.

In December 1962, he joined a select group of athletes who have been successful in their professional careers since their college football days, when he was chosen a member of the Sports Illustrated Silver Anniversary All-American team.

In August 1963, General Ryan was assigned to the Pentagon as Inspector General for the US Air Force. One year later he was named vice Commander in Chief of Strategic Air Command and in December 1964, became Commander in Chief. He was assigned as Commander in Chief, Pacific Air Forces, in February 1967.

Ryan was appointed vice Chief of Staff of the US Air Force in August 1968, and Chief of Staff of the US Air Force in August 1969. He commanded the Air Force during the height of the Vietnam War.

Gen Daniel "Chappie" James Jr.

- Gen Daniel "Chappie" James Jr. was a Tuskegee airman who flew one hundred combat missions during the Korean War.
- He flew over 60 combat missions as a full colonel during the Vietnam War.
- The first black in American history to attain the rank of four-star general.
- Served as commander of the North American Air Defense (NORAD) Command until his retirement in January of 1978.

One of the great Tuskegee airmen was Daniel "Chappie" James Jr. Chappie won his wings and a commission in 1943 but did not see combat in World War II. After the war, James quickly earned a reputation as an outstanding fighter pilot. In Korea he flew one hundred combat missions, and in Vietnam—by 1965 he was a full colonel—he flew over threescore more. Not only was that war unpopular, but racial unrest was exploding into violence all over the United States. James returned from Vietnam and was often called upon to defend not only America's military policies, but also its racial policies. An articulate speaker who commanded great physical presence (he was six feet, four inches and weighed nearly 250 pounds), he was an especially effective spokesman for the Air Force.

In 1967, he was named commander of Wheelus AFB in Libya just as Colonel Gadhafi succeeded in his revolution there. Gadhafi demanded that the air base, which he saw as a vestige of European colonialism, be closed and its facilities turned over to the Libyan people. This obviously was an extremely delicate position for James requiring restraint, tact, diplomacy, and grit. He displayed an abundance of all these qualities, and upon leaving Wheelus a year later, he received his first star.

After four years in the Pentagon working in Public Affairs where he won two more stars, he was named vice commander of Military Airlift Command (MAC). After less than two years at MAC, he was given a fourth star, the first black in American history to attain that rank, and was named commander of the North American Air Defense (NORAD) Command. Surprisingly for a man of his size and appearance, James was in poor health. He suffered a heart attack in 1977 and soon after elected to retire. His health continued to decline and in February 1978, one month after retirement, he suffered a fatal heart attack.

AIC John L. Levitow

On 24 February 1969, Airman Levitow flew combat patrol over South Vietnam as the loadmaster of an AC-47 Dragon ship. The gunship was patrolling in the vicinity of Tan Son Nhut when the Army post at nearby Long Sinh came under attack. The aircraft diverted to aid in the defense of the post. Firing its mini-guns at the enemy, the gunship knocked out two mortar positions, but further firings were observed a few kilometers away. As the AC-47 flew in the direction, a mortar shell fell on the top of its right wing. A brilliant explosion shook the aircraft violently and the fuselage was riddled by thousands of shell fragments. Airman Levitow and another crew member were standing near the open cargo door at that moment dropping parachute illumination flares. The explosion knocked both of them to the floor and a flare that they were handling was tossed inside the cargo compartment. Spewing toxic smoke, the activated magnesium flare was due to separate explosively from its cannister and ignite within seconds. Although stunned and wounded by shrapnel, Airman Levitow moved toward the flare and flung himself on top of it to keep the flare from rolling around the pitching cargo hold. He then dragged himself and the flare back toward the cargo door and tossed it out. The flare ignited just as it cleared the aircraft. Levitow was awarded the Medal of Honor for his selfless heroism that saved his fellow crew members and the gunship. President Richard M. Nixon presented the award at the White House on 14 May 1970.

An AC-47 at Nha Trang Air Base displays the three Gatling type machine guns which make up its armament. The AC-47s were the first gunships and were called a variety of nicknames—Puff, Spooky, and Dragon ship were among the most popular.

Gen George S. Brown

- As executive officer of the 93d Bomb Group, he took part in the famous low-level bombing raid against oil refineries at Ploesti, Romania, on 1 August 1943.
- In 1957 he served as executive to the Air Force Chief of Staff, Gen Thomas White.
- In June 1959 he was selected to be Military Assistant to the Deputy Secretary of Defense, and later was Military Assistant to the Secretary of Defense.
- As Seventh Air Force Commander in 1969, he was responsible for all Air Force combat air strike, air support, and air defense operations in Southeast Asia.

General Brown was born in Montclair, New Jersey, on 17 August 1918. He graduated from high school in Leavenworth, Kansas, and after attending the University of Missouri for a year, he received a congressional appointment to the US Military Academy, West Point, New York, in 1937. He graduated from the Academy in 1941, and entered flying training at Pine Bluff, Arkansas. He received his pilot wings at Kelly Field, Texas, in 1942.

In August 1942, he flew with the 93d Bombardment Group to England, the first B-24 group to join the Eighth Air Force. Until April 1944, he served in various positions with the group, including commander of the 329th Bombardment Squadron, group operations, and then executive officer. It was as executive officer that he took part in the famous low-level bombing raid against oil refineries at Ploesti, Romania, on 1 August 1943. The 93d Group was the second of five B-24 groups that raided Ploesti from a temporary base at Bengasi, Libya. The 93d Group, led by its commander, flew directly into heavy defenses to hit three of the six target refineries. The lead plane and 10 others were shot down or crashed on the target. General Brown,

then a major, took over the lead of the battered 93d and led it back to Bengasi. He received the Distinguished Service Cross for his actions on that mission.

During the Korean War in May 1952, he joined Fifth Air Force Headquarters at Seoul, Korea, as director for Operations. He served as the assistant to the chairman, Joint Chiefs of Staff, in Washington, D.C., from August 1966 to August 1968. He then assumed command of the Seventh Air Force and also became deputy commander for Air Operations, US Military Assistance Command, Vietnam (MACV). In his MACV position, he advised on all matters pertaining to tactical air support and coordinated the Republic of Vietnam and United States air operations in the MACV area of responsibility. In September 1970, General Brown assumed duty as commander, Air Force Systems Command, Andrews AFB, Maryland.

General Brown was appointed by the president to be Chief of Staff of the US Air Force, effective 1 August 1973, and to be chairman of the Joint Chiefs of Staff, Department of Defense, effective 1 July 1974. General Brown retired 21 June 1978, and passed away 5 December 1978.

Capt Richard S. "Steve" Ritchie

on. Ritchie fired two missiles but the MiG-21 was able to evade them. After countering the MiG's evasion "jinks," Ritchie fired two more air-to-air missiles. The MiG went into a thin overcast and when he came out, one of the missiles went by his left. Seeing the missile, he immediately broke to the right causing the second missile to impact and completely destroy the MiG. Ritchie's fifth kill broke the Vietnam record of four MiG kills set by Col Robin Olds in 1967.

On 28 August 1972 Capt Richard S. "Steve" Ritchie became the first USAF ace of the Vietnam War, downing his fifth MiG-21. Ritchie was a pilot attached to the 432d Tactical Reconnaissance Wing flying F-4 Phantoms. His fifth aerial victory came 30 miles west of Hanoi while he and his Weapon Systems Operator, Capt Chuck DeBellevue, were flying MiG Combat Air Patrol. After receiving information that "bandits" were threatening an F-4 strike force, Captain DeBellevue picked the MiGs up on radar. The MiGs were four thousand feet higher than Ritchie's F-4 requiring a quick climb and a hard turn to meet the bandits head

The MiG-21 was a light refined aircraft with Mach 2 capability. The proficiency of US aircrews mitigated the superior characteristics of the MiG-21 during the Vietnam War. Between 26 April 1965 and 8 January 1973, USAF F-4s and B-52s downed 68 MiG-21s.

Gen David C. Jones

- In combat, he was assigned to a bombardment squadron during the Korean War and accumulated more than three hundred hours on missions over North Korea.
- In 1969 he served in the Republic of Vietnam as deputy commander for Operations and then as vice commander of the Seventh Air Force.
- After serving four years as Chief of Staff of the Air Force, General Jones was appointed chairman of the Joint Chiefs of Staff on 21 June 1978.
- After retirement from the Air Force, he served on the board of directors for General Electric.

General Jones was born in Aberdeen, South Dakota. He graduated from high school in Minot, North Dakota, in 1939 and attended the University of North Dakota and Minot State College until the outbreak of World War II. He entered the Army Air Corps, beginning aviation cadet training in April 1942, and receiving his commission and pilot wings in February 1943.

In combat, General Jones was assigned to a bombardment squadron during the Korean War and accumulated more than three hundred hours on missions over North Korea. In 1969, he served in the Republic of Vietnam as deputy commander for Operations and then as vice commander of the Seventh Air Force.

General Jones' assignments included operational and command positions in strategic, tactical, and training units, as well as service in staff positions with major headquarters in the United States and overseas. He served as inspector, operator, planner, and Commander in Chief of United States Air Forces in Europe. Concurrent with duty as Commander in Chief USAFE, General Jones was commander of the Fourth Allied Tactical Air Force and led the way toward establishing the integrated air headquarters in NATO's Central Region, Allied Air Forces, Central Europe.

General Jones served four years as Chief of Staff of the US Air Force, responsible for administering, training, and equipping a worldwide organization of men and women employing the world's most advanced defense systems. Concurrently, he was a member of the Joint Chiefs of Staff.

General Jones was appointed Chairman of the Joint Chiefs of Staff, Department of Defense, on 21 June 1978.

Brig Gen Robin Olds

- Credited with 12 aerial victories in World War II and four in Vietnam making him a triple ace.
- Wingman on the first jet acrobatic team in the Air Force.
- Won second place in the Thompson Trophy Race (Jet Division) at Cleveland in 1946.
- Participated in the first one-day, dawn-to-dusk, transcontinental roundtrip flight in June 1946 from March Field, California, to Washington, D.C.

Robin Olds was born on 14 July 1922, in Honolulu, Hawaii, the son of an Army Air Corps major general. He graduated from the US Military Academy at West Point and commissioned as second lieutenant in June of 1943. While at West Point, he was honored for his football skills by being named as an All-American tackle in 1942.

He began his combat flying in a P-38 Lightning named *Scat I* during World War II and finished World War II flying *Scat VII*, a P-51 Mustang. During World War II, he flew 107 combat missions and was credited with 12 aerial victories and destroying 11 and one-half aircraft on the ground.

In February 1946, Olds started flying P-80 jets at March Field in the first squadron equipped with this jet. In 1948, he went to England as a Royal Air Force exchange pilot and flew the Gloster Meteor jet fighter in the No. 1 Fighter Squadron at RAF Tangmere.

Olds's most notable accomplishments occurred when he assumed duties as commander of the 8th Tactical Fighter Wing at Ubon Royal Thai Air Force Base, Thailand, in September 1966. While commanding this unit, he shot down two MiG-17s and two MiG-21s. Olds, the aircraft commander, and Lt Stephan B. Croker, the backseat pilot, scored two of those victories in a single day, 20 May 1967. Olds was a beloved leader who spent many hours talking to his junior officers constantly trying to improve his skills as well as theirs. On one such occasion, a junior officer recommended swapping an F-4's electronic countermeasure pods with an F-105's. The false signal would entice MiGs to engage the F-4's thinking they were engaging the slower, less agile F-105. Olds liked the idea and managed to cut through the red tape and employ the tactic. The resultant operation, code named Bolo, was a great success.

From Vietnam, Olds returned to the United States to serve as commandant of Cadets at the Air Force Academy. He retired from active duty in June of 1973. His decorations include: the Air Force Cross, Distinguished Service Medal, four Silver Stars, Legion of Merit, six Distinguished Flying Crosses, and 40 Air Medals. One of the truly legendary "stick and rudder" men of the Air Force.

Olds married film star Ella Raines.

Brig Gen Harry C. "Heinie" Aderholt

- Instrumental in developing Special Operations capabilities in USAF.
- Parachuted Korean spies behind communist lines in Korean War.
- Commanded airlift operations supporting Khamba revolt against China.
- Established system of Lima Sites in Laos.
- Commanded night, low-altitude, B-26 raids on Ho Chi Minh Trail.

Harry C. "Heinie" Aderholt was born in 1920 and joined the US Army Air Forces through the aviation cadet program in 1942. Commissioned out of pilot training in 1943, he served in North Africa and Italy as a B-17 and C-47 pilot until the end of the war.

During the Korean War General Aderholt began his long career in special operations by commanding a Special Air Warfare Detachment that parachuted Korean spies behind communist lines. Back in the United States he helped the Central Intelligence Agency with air operations, equipment, and training, before beginning his long series of assignments alternating between Asia and the US Air Force Special Air Warfare Center (predecessor of US Air Force Special Operations Command) in Florida.

In the early 1960s General Aderholt commanded the air operations supplying and assisting the Khamba tribesman during their guerrilla war against China in Tibet. He also developed the system of small airstrips in Laos (called Lima sites) that sup-

ported special operations and combat search and rescue operations in Laos and North Vietnam. In 1966 General Aderholt established the Joint Personnel Recovery Center in Saigon and then took command of the newly activated 56th Air Commando Wing in Thailand conducting devastating low-level night interdiction missions over the Ho Chi Minh Trail with old prop-driven aircraft. Later he served as chief of the Air Force Advisory Group in Thailand.

During his US tours, General Aderholt commanded the 1st Air Commando Wing and was instrumental in developing the Special Operations capabilities of the US Air Force. After retiring in 1972, he was called back to active duty in 1975 to command US Military Assistance Command in Thailand during the desperate fighting that brought the communists to power in Cambodia and Vietnam. In Thailand General Aderholt helped manage the evacuation of US personnel, employees, and allies, and ensured that US-Thai relations remained strong after Thailand's neighbors fell to the communists.

Rebuilding the Air Force
and the Gulf War

1976–91

El Dorado Canyon (14–15 April 1986)

The EF-111 Raven was used to jam Libyan search radar and provide electronic countermeasures against any surface-to-air threats.

On 14 April 1986 the United States launched an air attack, El Dorado Canyon, against targets in Libya. President Ronald W. Reagan declared the attack a reprisal for the Libyan role in the 5 April bombing of a West Berlin discotheque that killed US citizens. More than six hours before the attack, 24 F-111s took off from Lakenheath Air Base, Great Britain, escorted by 28 tankers and five EF-111 Ravens from Upper Heyford, England. The group represented the largest air-combat group assembled over England since World War II. The planes merged well away from land over the English Channel and began their twenty-eight hundred mile journey to Libya. Because the French and Spanish governments denied overflight privileges, the strike group was forced to fly a circuitous route around the Iberian Peninsula. Five hours after the F-111s took off, A-6 bombers and F-14s were launched from the carriers USS *America* and USS *Coral Sea*.

H hour was established as 7:00 P.M. eastern standard time. Ten minutes before H hour, Libyan radar operators reported confusing blips on their screens; the electronic countermeasures were already masking the attack. Minutes before the bomb runs began, missile attacks were launched against Libyan antiaircraft defenses. These proved to be effective. No MiG interceptors were ever launched and many sites turned off their electronic sensors to avoid being targeted by radar-guided high-speed antiradiation missiles. However, the Libyans did manage to put up some intense ground fire over Benghazi, including surface-to-air missiles and antiaircraft guns of all types.

Under the cover of darkness, 18 F-111 fighter-bombers roared across the coast of Libya at more than 500 MPH and only two hundred feet above the ground. They lined themselves up with an electronic beacon from a US base on Italy's Lampedusa Island. The American aircraft slipped in behind Libyan radar coming from the west and south to destroy the military side of the Tripoli International Airport, a naval facility, and Mu`ammar Gadhafi's barracks. Over Tripoli, the F-111s faced Soviet-made ZSU-23 four-barreled guns clustered around the Aziziya barracks. One F-111 was badly damaged by these guns. Capt Paul Lorence and Capt Fernando Ribas-Dominicci, weapon systems operator, tried to get the plane out to sea before ejecting. As they crossed the coast, however, their plane exploded killing both airmen. With the bomb run complete, the F-111s required eight more hours and two more in-flight refuelings in order to return to Lakenheath. The 15-hour mission was exhausting; some men had to be lifted out of their seats upon their arrival in Britain.

The operation achieved the objective of hitting terrorist infrastructure targets in Libya. Even with much publicity about mission shortcomings, the overall impact was one of success. At the Benina airfield, four MiG-23s, two Soviet helos, and two prop planes were destroyed. At the Tripoli International Airport, five Il-76s and several buildings were destroyed. The Sidi Bilal naval base, as well as military barracks in Benghazi and Tripoli were damaged. The United States was able to achieve total surprise, evidenced by lack of air raid alarms being sounded and street lights illuminating the targets. The complex operation involved 150 aircraft and resulted in the dropping of 60 tons of bombs. The chairman of the Joint Chiefs of Staff, Adm William Crowe, declared the operation "very successful" and made the world smaller for the terrorists.

Operation Just Cause (December 1989)

The AC-130 Spectre gunship.

By 1989 the administration of President George W. Bush decided that the Panamanian dictator Manual Noriega had to be ousted from power because of his involvement in the narcotics trade and his developing rapprochement with communist Cuba. After the failure of various political initiatives, Bush decided upon the use of force. The result was Operation Just Cause, which began on the night of 19 December 1989.

The Air Force played a central role in Just Cause. No component was more deeply involved than the Military Airlift Command (MAC). In the weeks before the invasion, MAC helped to pre-position troops and equipment for Just Cause by ferrying them to Howard Air Force Base, Panama. The plans for Just Cause also sought tactical surprise through airborne assaults by units flown directly from the United States. MAC managed to airlift three companies of the 75th Ranger Regiment and one brigade of the 82d Airborne Division for an assault on Torrijos-Tocumen Airport near Panama City. Forty-five hundred paratroopers dropped in the dead of the night, the largest airborne assault since World War II. MAC transports also brought five companies of the Army's 75th Rangers from Lawson Army Air Field, Georgia, to Rio Hato, Panama, where the rangers were to engage and destroy the 6th and 7th Companies of the Panamanian Defense Forces—both mainstays of Noriega's misrule. In all, MAC employed 86 aircraft—63 C-141, 21 C-130s,

and 2 C-5s. Strategic Air Command (SAC) also provided RC-135s and U-2s which patrolled between the Yucatan Peninsula and Cuba to determine if the Cubans had provided Noriega with advance warning. They had not. Apparently the transports successfully avoided Cuban radar by descending to low altitude as they passed by Cuba.

By the time the transports reached their objectives in the early hours of 20 December, fighting had already begun. A task force from the 193d Infantry Brigade was driving from the Panama Canal Zone toward the headquarters of Panamanian Defense Forces. Two AC-130 gunships opened the attack on the headquarters. Their weapons, however, could not penetrate the concrete floors of the building, which permitted defenders to sally out to engage the approaching infantry task force. In the resulting firefight, the neighborhood of El Chorillo caught fire, killing several hundred civilians. Two F-117s also attacked Rio Hato, dropping bombs near military barracks. Alerted by explosions, the Panamanians quickly brought the parachuting Rangers under fire. It took two AC-130s to suppress the defenders with automatic weapons, certainly preventing a disaster. At Rio Hato—and elsewhere in Panama—the AC-130 showed itself a superb instrument for providing highly accurate fire support for ground forces, as well as night surveillance and navigation assistance.

During the invasion, Noriega fled to the residence of the Vatican's ambassador, where he surrendered on 3 January 1990. Fighting had effectively ended by Christmas day, though Just Cause did not officially end until 4 February 1990. A few figures convey the scale of the operation: the Air Force flew 40,000 passengers to and from Panama and delivered nearly 21,000 tons of cargo. There were no accidents, and no airman lost his life. Just Cause, while not unmarred by errors, was on the whole a highly successful limited operation in support of political objectives.

Strategic Attack on Iraq (January 1991)

Initial air attacks of Instant Thunder.

The first phase of Desert Storm was the strategic air campaign, initially code named Instant Thunder. Gen Buster Glosson reworked the plan developed by Col John Warden and the Pentagon Checkmate staff so that it addressed Iraqi forces in Kuwait as well as command structures. Strategic attacks on Iraq focused on three "centers of gravity": Iraqi National Command Authority; chemical, biological, and nuclear capability; and the Republican Guard Forces Command.

Air superiority was the first priority of coalition forces and was achieved in the first days of the campaign. The Gulf War was the first conflict in history in which a large percentage of the air-to-air engagements that produced kills involved "beyond visual range" shots. Coalition counterair operations were so effective that the Iraqi air force lost more than 400 combat aircraft while only one coalition plane was shot down by an Iraqi plane.

Coalition destruction of Iraqi ground-based air defenses was equally as overwhelming. The best measure of its effectiveness was the relative immunity with which coalition aircraft operated over Iraq. During the first four days of Desert Storm, radar SAMs downed or damaged nine coalition fixed-wing aircraft. For the remainder of the campaign, radar SAMs hit only four more coalition aircraft in spite of the fact that coalition air forces maintained a daily average of over

1,600 "shooter" sorties and another 540 combat-support sorties.

Precision strikes in the first two nights focused on the Iraqi leadership and its command, control and communications (C^3). Because of the heavy defenses, physical hardness, small size and urban location of these targets, the F-117/GBU-27 combination was the primary attack weapon systems. There were over 70 strikes carried out on 17–18 January against leadership and C^3 targets by F-117s, tactical land attack missiles, conventional air launched cruise missiles, and F-111Fs. Hits from 2,000

The 38-day strategic air campaign was so effective that the ground campaign took only 100 hours

---

pound bombs within feet of desired aiming points on government ministries, national command and control facilities, headquarters and communications centers destroyed or severely disrupted Iraqi C^3. The breakdown in Iraqi C^3 was so complete that at the cease-fire meetings that ended the war, coalition commanders had to tell the Iraqi generals where the Iraqi troops were.

Coalition airpower was overwhelming in both numbers and in quality. Coalition fighters, attack aircraft, and airlift were superior and in conjunction with coalition aerial refueling, reconnaissance, electronic warfare and airborne command and control capabilities, they gave the coalition air dominance from the early days of the campaign.

The F-117 Nighthawk was spectacularly successful against even the toughest targets.

The Battle of Khafji (1991)

Destroyed Iraqi armored columns.

On 17 January 1991 coalition airpower started offensive actions against military targets in Iraq and Kuwait. Planning depended upon the air campaign to attrit Iraqi forces, prevent them from detecting coalition redeployments to the west, and to lock Iraqi forces in place.

Indeed, Iraqi forces were well prepared for a frontal ground assault against their positions in Kuwait. During months of preparation, miles of fortified positions, obstacles, minefields, and fire trenches were constructed just north of the Saudi-Kuwaiti border. Saddam Hussein was convinced that bloody ground action would be the ultimate determining factor. During an interview, he declared that the United States relied on its Air Force, but that air forces have never been decisive in war.

Neither the Iraqis, nor the world in general, expected independent air operations past the first week. Indeed, classical theory and exercise practice only allotted five to seven days for independent air actions in combined arms operations. The secretary of defense and chairman of the Joint Chiefs stated on 23 January that air supremacy had been achieved and there was "no hurry" to stop pounding the Iraqi forces.

By the second week, Iraqi commanders realized their much touted air defense system was not exacting the toll they had counted on. They further realized air action was steadily destroying their forces and the means of support and sustainment, as well as the means of command and control. They were facing a "use it or lose it" situation and needed to recapture the initiative.

Saddam started ground actions on the night of 29 January to spur the coalition into the bloody ground war of attrition he desperately needed for either military or political victory. The Iraqi 5th Mechanized Division units attacked across the Saudi-Kuwaiti border at three locations. Their 1st Mechanized Division was to conduct screening operations to the west and their 3d Armor Division was to exploit any breakthroughs. Many other Iraqi units also had supporting roles and were prepared to "grind down" any subsequent coalition ground penetrations into Kuwait.

Two prongs of the attack were stopped by US Marine Corps units with close air support. One Iraqi battalion did make it into the town of Khafji that had been evacuated because it was within range of Iraqi artillery in Kuwait. The joint surveillance, target attack radar system pinpointed the movements of Iraqi follow-on reinforcements for airpower strikes. With airpower halting the majority of Iraqi forces in Kuwait, the Saudis reclaimed the town of Khafji on 31 January.

While combat activity in the town was not extensive, movement of the reinforcing divisions, stopped by airpower, was substantial. The Joint Force Commander's refusal to take Saddam's bait and the predominant use of airpower allowed redeployments and preparations for the "Hail Mary" to continue without interruption. The impact on the Iraqis was even more momentous as they attempted no other offensive. Indeed, subsequent activity to dig deeper, disperse, use more deception, move frequently, and use smaller convoys demonstrated renewed efforts to simply survive continuing joint air attacks.

Col John W. Warden, III

- Author of *The Air War.*
- Created the "Five Rings" model of enemy systems and "Inside-out" warfare.
- Developed the original draft for the "Instant Thunder" plan for the air campaign in the Gulf War.
- Reinvigorated the Air Command and Staff College and Airpower Theory throughout the Air Force.

John Warden had a full operational career including 266 combat missions in Vietnam as Forward Air Control pilot flying OV-10s and flying and command assignments in F-4 and F-15C units culminating in command of a F-15C Fighter Wing. He is best known, however, as one of the leading airpower theorists of the late twentieth-century and as the guiding light behind the Gulf War air campaign.

Colonel Warden's extensive writings contain many original, provocative, and influential ideas and he continues to be a prolific author and speaker. One of his simplest and most influential ideas is that the enemy (whether a nation or a drug cartel) can be thought of as a system consisting of five concentric rings: leadership, system essentials, infrastructure, population, and fielded military forces. The most important ring, leadership, is at the center and fielded military forces are on the outside protecting all the others (see figure). Airpower is uniquely capable of attacking any of these rings and is most effective when used against the most important inner rings rather than the less important outer rings. Attacking the inner rings and then working outward

is sometimes called "inside-out" warfare. This idea was at the core of the air plan Warden and his subordinates on the Air Staff drafted for the Gulf War. The plan as ultimately executed was enormously successful in paralyzing the Iraqi leadership and infrastructure before moving on to cripple the Iraqi ground forces which were finished off by the ground invasion.

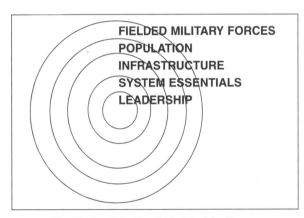

The Warden "Five Ring" Model of the Enemy.

Gen Charles A. Horner

- As JFACC for Operation Desert Shield/Storm he commanded all air operations in the Gulf War.
- Flew 112 F-105 combat missions over North Vietnam during the Vietnam War.

Gen Charles Horner received his Air Force commission in 1958 from the Reserve Office Training Corps (ROTC) program at the University of Iowa. During the Vietnam War he flew 41 combat missions over North Vietnam in F-105 fighters and an additional 71 combat missions in F-105 Wild Weasel aircraft, hunting down North Vietnamese air defenses. During his distinguished operational career he commanded a tactical training wing, a fighter wing, two air divisions, a numbered Air Force, and served as commander in chief of the North American Aerospace Defense Command and US Space Command. He is best known for his five years as commander of 9th US Air Force and US Central Command Air Forces (1987–92) and particularly his command of air operations during the Gulf War (1991).

During the Gulf War General Horner served as joint force air component commander (JFACC) commanding all coalition air operations. In this capacity he managed the enormously complicated air portion of Operation Desert Storm, employing more than 2,600 aircraft from 11 countries. General Horner's leadership helped produce one of the most rapid and devastating air campaigns in military history. This campaign not only wiped out the Iraqi air force and air defenses but also destroyed some of the Iraqi infrastructure for building chemical, biological, and nuclear weapons, and large parts of the Iraqi army. The campaign disrupted Iraqi command and control so effectively that at the surrender negotiations, the US representatives had to tell the Iraqi generals where the Iraqi troops were. Most impressively, Horner accomplished all this in just over 40 days at a cost of only 42 coalition aircraft against very powerful and experienced Iraqi forces.

After his retirement in 1994, General Horner has lectured, consulted, and written extensively on defense matters including a book on the Gulf War, *Every Man a Tiger*, which he coauthored with Tom Clancy.

Gen Merrill A. McPeak

- Radically reorganized the USAF to meet new post-Cold War challenges.
- USAF Chief of Staff during the Gulf War.
- Leading advocate of the "Composite Wing."

Merrill McPeak was commissioned in 1957 from the ROTC program at San Diego State College. He was a demonstration pilot in the US Air Force Air Demonstration Squadron (Thunderbirds) for two years before going to Vietnam where he fought as an attack pilot and a forward air controller. He went on to command a wing, a numbered Air Force, and Pacific Air Force before being named Chief of Staff of the US Air Force in late 1990.

Appointed unexpectedly on the eve of the Gulf War, General McPeak immediately guided the Air Force through Operations Desert Shield and Desert Storm, the largest airlift and largest air war in decades.

After the triumph of the Gulf War, General McPeak became perhaps the most controversial Chief of Staff in Air Force history. He pushed the Air Force through the most extensive reorganizations it

had ever experienced. The most visible change was that he scrapped the old three-part Air Force structure of Tactical Air Command (TAC), Strategic Air Command (SAC), and Military Airlift Command (MAC). The new Air Force organization fit better with the Goldwater-Nichols Defense Reorganization Act and was the critical first step in reshaping the Air Force to meet the needs of the post–Cold War era. He was also a leading advocate for the Composite Wing concept that combined several different aircraft types in a single air wing.

Not everyone welcomed the changes General McPeak made in the Air Force, but he was not deterred by criticism or opposition from following the path he felt was best for the service. Oddly enough, he received more vociferous criticism for his ill-fated efforts to change Air Force uniforms than for the enormous changes he made in the way we do business.

95

Post-Cold War

1992–2000

Policing Postwar Iraq (1992–?)

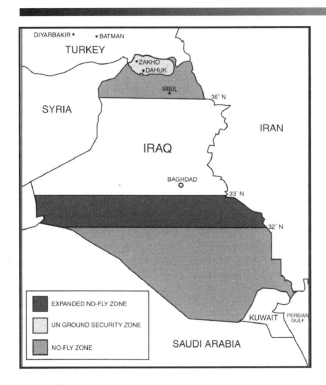

The Gulf War drove Saddam Hussein's forces out of Kuwait and weakened him dramatically and this prompted rebellions in March 1991 by ethnic Kurds in northern Iraq and the Shiite religious group in southern Iraq. The rebels, however, were not well equipped and the international community did not support their efforts to break away from Iraq because that would have further destabilized the already unstable Middle East. Without international military support the rebels were too weak to face the Iraqi army and they were soon defeated.

The defeat of the Kurdish forces in the north created a massive refugee problem as more than a million Kurds fled their homes to escape violent reprisals by the Iraqi army. The United States and the United Nations responded to this humanitarian crisis with Operation Provide Comfort in April 1991. In order to stabilize the situation, US and coalition forces launched an airlift to deliver relief supplies and used ground forces to establish a ground security zone in northern Iraq and refugee camps in northern Iraq and southern Turkey to facilitate distribu-

tion of supplies. What made these efforts possible was coalition air supremacy.

The ground security zone (where no Iraqi troops were allowed) and the "no-fly" zone (where no Iraqi aircraft were allowed) made it safe for the Kurds to return to their homes and by the end of May almost all of the refugees had returned and by mid-July the coalition ground forces had withdrawn from Iraq. The United States has continued to maintain the no-fly zone ever since and in recognition of the end of the transition from a humanitarian mission to one of monitoring Iraqi airspace, Operation Provide Comfort was replaced by Operation Northern Watch at the beginning of 1997.

Shortly after the Gulf War, the Iraqi army put down a rebellion by Shiite Moslems in southern Iraq and the repression there was so severe that the United Nations adopted a resolution to protect them from Iraqi air attack. In August 1992 the United States announced a no-fly zone over southern Iraq. Maintenance of the southern no-fly zone has been the task of Operation Southern Watch ever since. In October 1994, in response to Iraqi troop movements that threatened another invasion of Kuwait, the United States declared the southern no-fly zone a no-fly/no-drive zone. In 1996, in response to renewed Iraqi attacks on the Kurds, the United States expanded the southern no-fly zone and launched extensive attacks (Operation Desert Strike) to destroy Iraqi air defenses in the new patrol areas.

Since the completion of the airlift and humanitarian relief phases of Operation Provide Comfort, US and coalition efforts have focused on continuous intelligence gathering, surveillance, and reconnaissance over Iraq. These efforts have put a heavy strain on E-3, RC-135, and other surveillance aircraft and units but have also produced some dramatic successes. In counterair operations the most notable victories were the downing of an Iraqi MiG-25 in December of 1992 by a US F-16 assigned to Southern Watch and the downing of an Iraqi

MiG-29 in January 1993 by a US F-16 assigned to Provide Comfort.

The most dramatic impact of operations has been in strategic attack. When the Iraqi's continued to block UN inspectors trying to dismantle Iraq's missile and weapons of mass destruction (WMD) programs, the United States and United Kingdom launched a four-night series of attacks against roughly 100 strategic military targets in Iraq. These attacks in December 1998 (Operation Desert Fox) struck Iraq's military through destruction of air defense, command and control facilities, and air bases. Oil facilities used by Iraq to evade UN economic sanctions were also attacked. Most importantly, though the Iraqis could keep inspectors out of their missile and WMD sites, they could not defend the sites from air and missile attacks, so Desert Fox shut them down. In addition to F-15, F-16, F-117, A-10, and B-52, the strike missions during Desert Fox witnessed the combat debut of the B-1B.

Though the Iraqis have not shot down any coalition aircraft during our operations against Iraq after the Gulf War, these missions have not been cost-free. Two tragedies (the accidental shoot-down of two US Army helicopters over northern Iraq by USAF fighters and the death of 19 US airmen in a terrorist attack in Saudi Arabia) have reminded us of the difficulties and dangers of these operations. Both of these events have led to improvements in US operations to prevent a repetition. The demands of 10 years of operations against Iraq and the end of the Cold War have led to a major reorganization of the US Air Force into Aerospace Expeditionary Forces.

Gen Ronald R. Fogleman, then Air Force Chief of Staff, summed up our postwar operations over Iraq nicely when he said that "What we have effectively done since 1992 is conduct an air occupation of a country."

The B-1B bomber made its combat debut against Iraq in December 1999 as part of Operation Desert Fox.

Operations in Somalia (1992–94)

An AFSOC MH-53 Pave Low delivers relief supplies to a remote part of Somalia.

Two years of civil war tore apart Somalia in the early 1990s and contributed to a famine that killed about one-third of a million people in 1992 and drove three-fourths of a million refugees out of the country. A shaky cease-fire went into effect in March 1992, United Nations troops from Pakistan arrived in July, and the United States launched Operation Provide Relief in August. By late December, USAF C-141s and C-130s had delivered over 19,000 tons of food to starving Somalis. Unfortunately, armed Somali groups were intimidating relief workers and stealing the food supplies. They even shelled cargo ships delivering supplies, and C-130s were hit by ground fire. Additional UN troops were airlifted in, but when the situation did not improve, the United States launched Operation Restore Hope in December 1992.

Operation Restore Hope deployed overwhelming ground forces from 23 countries into Somalia and secured the ports, airfields, and lines of communication necessary to conduct effective relief operations. In addition, Marine and Navy forces came in by sea, and the Restore Hope airlift brought more than 32,000 troops and 32,000 tons of supplies into Somalia. This massive operation required more than 1,100 aerial refueling sorties by KC-135 aircraft. Operation Provide Relief ended in February 1993 and Restore Hope ended in May with the turnover of relief operations to the United Nations and the redeployment of the vast majority of the 25,000 US troops that had been brought into Somalia.

Unfortunately, the UN's efforts to build a stable government in Somalia soon brought it into conflict with a Somali warlord (Mohammed Farah Aidid) who felt he was not getting as much power as he deserved. In June, Aidid's forces ambushed and wiped out a Pakistani convoy and from then on the UN was in open conflict with Aidid. US Air Force and joint special operations units were sent to try to capture Aidid but his supporters kept him hidden. The US forces conducted raids to capture his closest subordinates, as a means of increasing pressure on Aidid. On 3 October 1993 US forces conducted a successful raid deep into Aidid's stronghold and captured some of his key supporters, but the Somalis managed to shoot down one of the US Army helicopters. The ensuing combat search and rescue effort turned into a major pitched battle in the streets of Mogadishu. When the battle ended the next day, 19 US troops and hundreds of Somalis were dead. The United States immediately launched Operation Restore Hope II which airlifted US combat forces back into Mogadishu (including 18 M-1 tanks and 44 Bradley fighting vehicles) and returned AC-130s to bases in Kenya from which they could attack targets in Somalia. These 8,000 mile one-way flights directly from Georgia to Mogadishu were another airlift triumph, but the willingness of thousands of Somalis to die for Aidid indicated that the UN's plan to leave him out of the future government was doomed and the last US troops left in March 1994.

Operations Provide Promise and Deny Flight: Bosnia (1992–96)

C-130s, like this one being deiced before a mission in the Balkans, were the backbone of Operation Provide Promise.

Serbs were the largest ethnic group within Yugoslavia and had most of the political power, but this caused only limited friction within the country until 1990 when the collapse of the Soviet Union ended the threat of foreign invasion and the communist party began to lose control of the country. Without communism and foreign threats, ethnic and religious loyalties became the strongest bonds among the people, and the various republics within Yugoslavia began to break away and declare independence. Slovenia and Croatia declared independence in 1991 and Bosnia did so in 1992. The Serb-dominated government of Yugoslavia, under Slobodan Milosevic, opposed the independence of the former Yugoslav republics as did the Serbian minority in the break-away republics.

With Milosevic's support, the heavily armed Serb minority in Bosnia began a campaign of murder, rape, and arson that became known as "ethnic cleansing." The purpose of this campaign was to terrorize the Bosnian Croats and Muslims into fleeing large parts of Bosnia that the Serbs would then occupy and separate from Bosnia. The ethnic cleansing campaign killed thousands and left millions homeless. The United States began to airlift humanitarian supplies into Bosnia in July 1992 as part of the UN airlift. The US operation was named Provide

Promise and continued until January 1996. The core of the airlift operations were deliveries of aid to Sarajevo, the besieged capital of Bosnia, where nearly 400,000 people depended on Provide Promise for three and one-half years. During that time aircraft from the United States and 20 other nations flew nearly 13,000 sorties, delivered more than 160,000 tons of food, medicines and supplies, and evacuated 1,300 wounded Bosnians. In addition to being the longest humanitarian airlift operation in USAF history, Provide Promise was one of the most hazardous. Ten US aircraft were hit by hostile ground fire and an Italian aircraft was shot down with the loss of all four crewmen.

Sarajevo was not the only besieged Bosnian city in danger of being ethnically cleansed. Several other cities were surrounded by hostile Serbs who periodically cut off ground contact with the outside world. US aircraft had to air-drop supplies into these smaller cities that did not have airports. The danger of Serbian ground fire was so severe that some of these loads had to be dropped from altitudes over 10,000 feet and numerous innovations were required to enable even GPS-equipped C-130s to successfully complete more than 2,200 airdrop sorties. While C-130s were the backbone of the Provide Promise airlift, USAF C-141s, huge C-5s, and new C-17s also participated.

The United States and the UN were not content to merely relieve the suffering, they also took positive steps to try to stop the fighting and punish the aggressors. With this in mind, the UN Security Council authorized NATO to enforce a no-fly zone over Bosnia. This enforcement operation (Deny Flight) began in April 1993. For example, in February 1994, Capt Robert L. Wright (an F-16 pilot with the 512th Fighter Squadron) shot down three Serbian jets in a single day. Unfortunately, the politics of enforcing the UN resolutions became very complicated because UN peacekeepers were deployed in Bosnia to monitor one of the early cease-fires. These forces were scattered throughout the country in small, lightly armed groups ideal for monitoring a cease-fire but unable to defend themselves effectively from Serb attacks. Whenever US and NATO airstrikes began to hurt them, the Serbs simply took peacekeepers hostage and forced a halt to the bombing.

By the summer of 1995 NATO had run out of patience with the Serbs and a mortar attack on the Sarajevo marketplace was the last straw. In July and August the peacekeeping forces on the ground were reinforced and consolidated so that they could defend themselves against Serb attacks. On 30 August, NATO air forces (under the command of Lt Gen Michael Ryan) began an aggressive strategic attack and counterland campaign called Operation Deliberate Force that made full use of precision guided munitions. At the same time, Muslim and Croat ground forces launched a major ground offensive against the Serbs. The Serbs quickly capitulated, so attacks stopped on 14 September. The operation officially ended on 21 September. Peace talks in November at Wright-Patterson AFB, Ohio, produced the agreement that ended the long war in Bosnia, and Operation Deny Flight ended in December with Operation Provide Promise ending the next month, in January 1996.

US airpower was not the only thing that finally brought a just peace to Bosnia but it was decisive and essential to feeding the hungry, protecting the innocent, and punishing the aggressors.

NATO counterair operations drove Serbian aircraft from the sky during Operation Deny Flight and strategic attack and counterland strikes during Operation Deliberate Force drove the Serbs to sign the peace treaty that ended the war in Bosnia.

Operation Uphold Democracy: Haiti (1994)

Unloading a C-5 in Haiti.

In December 1990, a leftist priest named Jean-Bertrand Aristide was elected President of Haiti by a landslide. He took office in February 1991 but was removed by a military coup in September. The United States, the Organization of American States, and the United Nations all opposed this assault on democracy and instituted diplomatic and economic sanctions. The vast majority of Haitians still supported Aristide and the new government could only stay in power through massive repression.

As the Haitian economy collapsed, the country was wracked by violence and a growing number of Haitians tried to flee the 700 miles to the United States in small boats. Vigorous diplomatic efforts, including an agreement between President Aristide and the military government in mid-1993, failed to resolve the crisis. Finally in September 1994 the US launched Operation Uphold Democracy.

US Atlantic Command had prepared two plans for the operation. If the Haitian military resisted, an airborne assault (like Operation Just Cause in Panama in 1989) would forcibly remove the current government and take over the country before handing it over to Aristide, the elected president. If the Haitian military capitulated, the United States would conduct an unopposed occupation of Haiti to facilitate a peaceful transition of power. The US Air Force had already launched the first two waves of aircraft (a total of 60 C-130s) loaded with US Army Airborne troops and equipment rigged for airdrop when the military government agreed to step down. The military government seemed to have realized, either through news reports or some other means, that the US had launched an invasion force and was suddenly anxious to back down before the fighting started.

The sudden capitulation of the illegal Haitian government demonstrates that an aggressive military plan, well resourced, and vigorously executed, can intimidate an adversary into capitulation even before the first shot is fired. The quick capitulation prevented a good deal of bloodshed but enormously complicated the execution of Uphold Democracy because the airlift had to switch from the invasion plan to the permissive-entry plan in midstream. When the US Joint Chiefs of Staff (JCS) got word that the Haitian military had agreed to step down, they ordered the aircraft already on their way to Haiti to turn around and return to base, a 24-hour pause in the movement to Haiti; and ordered the execution of the permissive-entry plan instead.

The Kosovo Crisis (1999)

Refugees from Kosovo at a NATO refugee camp in Albania, April 1999.

In the late 1990s Serbian dictator Slobodan Milosevic directed violent repression against the ethnic Albanian population in the Serbian province of Kosovo. After several brutal massacres of ethnic Albanians by Serbian forces, NATO demanded that Milosevic allow NATO peacekeepers to police Kosovo. When Milosevic refused, NATO launched an air campaign to force him to comply.

NATO ruled out ground combat, so from the outset on 23 March 1999, the NATO air campaign (Operation Allied Force) was the only combat effort. The initial hope was that Milosevic was bluffing and that after a few days of air attacks, he would have the excuse he needed to give in to NATO demands, remove his troops from Kosovo, and allow NATO troops to move in. Instead he used the air campaign as an excuse to dramatically increase the scale of his attacks on the ethnic Albanian population in Kosovo. Within a week of opening the air campaign, Milosevic's troops had killed thousands of Albanian civilians and driven nearly one million from their homes.

The vast and sudden flow of refugees out of Kosovo and into neighboring Albania and Macedonia created a massive humanitarian crisis and the United States responded with a huge humanitarian relief Operation: Shining Hope. During its first 50 days, Shining Hope forces (under the command of Commander 3d US Air Force, Maj Gen William S. Hinton) delivered more than 3,400 tons of food, equipment, and medical supplies to the refugees. The ability of the United States and NATO to sustain the Kosovo refugees in Albania and Macedonia was critical to our ultimate success in returning them to their homes and thwarting Milosevic's attempt to wipeout the ethnic Albanian population in Kosovo.

The Joint Forces Air Component Commander (JFACC) for Allied Force, Lt Gen Michael C. Short, had always favored a more aggressive air campaign. When the initial small-scale air attacks failed to convince Milosevic to comply with NATO demands, General Short got his wish. The United States and our NATO allies dramatically increased the forces committed to operations against Serbia. The political leaders also approved additional targets for strategic attack while NATO commenced a major counterland campaign to punish the Serb force in Kosovo. This sort of gradual escalation was not the most efficient way to use airpower, but was ultimately successful in achieving NATO goals after a 78-day campaign and contributed to Milosevic's downfall the following year.

By the time Milosevic capitulated, NATO air forces had conducted over 38,000 sorties. NATO counterair operations rapidly wiped out the Serbian air force but Serbian air defenses continued to pose a significant threat throughout the 11-week campaign. Learning from the rapid destruction of Saddam Hussein's air defenses during Desert Storm, the Serbs did not

NATO troops in Bosnia inspect the remains of a Serbian MiG-29, shot down by allied airmen.

immediately and aggressively try to defend their airspace. Instead, they kept their air defenses moving or hid them in populated areas (where collateral damage considerations would keep NATO from attacking them) and tried to ambush NATO aircraft when they got the opportunity. NATO tactics overcame this sort of guerrilla-warfare technique of air defense. Only two NATO aircraft were shot down during Operation Allied Force and both pilots were rescued by bold Combat Search and Rescue operations. This new air defense tactic did, however, force NATO to consider the surviving air defenses throughout the campaign.

In keeping with NATO's humanitarian goals, the entire Allied Force air campaign was conducted under very tight collateral damage restrictions. To achieve the unprecedented level of precision called for by these rules, the campaign relied more heavily on precision guided munitions (PGM) than Desert Storm. In all, Serbian targets were attacked with over 23,000 NATO bombs and missiles but there were only 20 incidents of collateral damage. In spite of this spectacular NATO success, the Serbs conducted effective information operations to make sure that the few accidental civilian casualties received a great deal of public attention. The diplomatic difficulties created by collateral damage and the unfortunate chain of events that led to an attack on the Chinese embassy in Belgrade reinforced the need to continue to improve the Allied targeting process and the precision of our attacks.

The counterland campaign was particularly challenging for NATO airmen. The rugged wooded terrain in Kosovo and the persistent

The B-2 and the JDAM made their combat debut during the Kosovo crisis and were enormously effective.

bad weather made Serbian ground forces very hard to find from the air. These difficulties were compounded by the fact that the Serbs did not have to face NATO ground forces so they could concentrate on hiding from NATO aircraft. In the last three or four weeks of the campaign, new tactics and improved weather combined to dramatically increase Serbian losses and helped force Milosevic's surrender.

Two systems that made their combat debut during Allied Force were the B-2 bomber and the GBU-31 joint direct attack munition (JDAM). The JDAM was still in testing when Operation Allied Force began but was used extensively and achieved amazing accuracy under all weather conditions. B-2 "stealth" bombers made 30-hour round-trip missions from their home base in Missouri, each dropping up to 16 2,000 pound JDAM's on up to 16 different targets with devastating results.

Gen Ronald R. Fogleman

- As Chief of Staff of the USAF he renewed service commitment to high technology and space.
- Flew 315 combat missions during the Vietnam War.
- Shot down over South Vietnam and rescued, under fire, by a US helicopter.

Gen Ronald Fogleman was commissioned out of the Air Force Academy in 1963 (the Academy's third graduating class). He flew 315 combat missions over Vietnam in F-100 aircraft as a fighter pilot and high-speed forward air controller. On one of these missions he was shot down and was rescued from a rice-paddy by a US Army helicopter as the Vietcong were closing in on him. He was wounded by Vietcong fire as he clung to the skid of the damaged US helicopter.

Later in his career, General Fogleman earned a masters degree in military history and political science at Duke University, taught history at the Air Force Academy, flew as an F-15 demonstration pilot for international air shows, and commanded an Air Force wing, an air division, and a numbered air force. He also served in Korea as deputy commander in chief, United Nations Command; deputy commander, US Forces Korea; and commander Republic of Korea/US Air Component Command before becoming commander in chief of US Transportation Command.

General Fogleman served as Chief of Staff of the US Air Force from 1994–97 and tried to calm some of the passions that his predecessor aroused. General Fogleman's most visible change was "stylistic." He made sure that he spent as much time as possible out with the force and he truly listened to the concerns of airmen out in USAF units all over the world. General Fogleman stressed personal accountability and renewed the Air Force's commitment to its core values: integrity, service before self, and excellence in all we do. Two of his primary areas of concern were high technology and the role of the Air Force in space. As Chief of Staff he did a lot to start the Air Force on its transformation into an Air and Space Force and to help the entire organization recognize that we will ultimately become a Space and Air Force.

Lt Gen Michael C. Short

- As JFACC during Operation Allied Force, he is credited with being the commander of the only air campaign to win a war without the assistance of friendly surface forces.
- Flew 276 combat missions during the Vietnam War.

Gen Michael Short was commissioned out of the Air Force Academy in 1965. He flew 276 combat missions as an F-4 pilot in the Vietnam War and commanded two squadrons and the F-117 "stealth fighter" group (4450th Tactical Group) while it was still a classified program. He also commanded three wings and a numbered Air Force. He is best known for his work as joint force air component commander (JFACC), commanding NATO air operations during Operation Allied Force (the conflict over Kosovo).

Operation Allied Force has been described as the first time in history that a war has been fought and won purely by airpower without the assistance of friendly surface forces. The United States and NATO Air Forces were not designed to fight without the involvement of friendly ground troops and US Air Force leaders did not advocate an all-air strategy, but that was the approach favored by NATO's political

leadership and General Short made it work. That was no easy task.

General Short had to work within very strict political guidance coming from not just US policy makers but a wide variety of NATO capitals as well. The initial plan had envisioned only a small, brief bombing campaign like the earlier one in Bosnia. When Serbian dictator Slobodan Milosevic continued to resist NATO demands, the forces involved had to be dramatically expanded to more than three times the initial force and nearly three times as many bases in a total of 15 countries.

Victory over Milosevic ultimately required 78 days and over 38,000 sorties. General Short led this air campaign so effectively that only two manned NATO aircraft were lost and no NATO troops were killed or injured by enemy fire.

Since his retirement in 2000, General Short has lectured widely on defense topics.

Part II

Aerospace Craft
of the
United States Air Force

Early Aircraft

Wright 1909 Military Flyer

Top Speed: 40 MPH
Range: 125 mi
Endurance: one hour, 12 min,
         40 secs

The world's first military airplane. It was built in response to United States Army Signal Corps specification 486, issued December 1907. The specifications for the aircraft were a speed of 40 miles per hour (MPH), a range of 125 miles, "and the ability to steer in all directions without difficulty." On 27 July 1909, Orville Wright, with Lt Frank P. Lahm as passenger, flew the aircraft for one hour, 12 minutes, 40 seconds and covered 40 miles, which met the Army's endurance requirements. Three days later, Orville Wright, with Lt Benjamin D. Foulois as passenger, covered a 10-mile test course at an average speed of 42 MPH, earning the Wrights a performance bonus (10 percent of the aircraft's base price for each MPH over 40). The aircraft was accepted on 2 August 1909 and redesignated Signal Corps Aeroplane Number 1.

Curtiss JN-4D "Jenny" (1916)

Top Speed: 75 MPH
Ceiling: 6,500 ft
Engine: Curtiss OX-5, liquid-
      cooled V-8 of 90 hp
Endurance: two hours, 30 min

The Jenny was generally used for primary flight training, but some were equipped with machine guns and bomb racks for advanced training. After World War I, hundreds were sold on the civilian market. The "Barnstormer" airplanes of the 1920s attracted many future World War II air force leaders to the thrills of flying. Powered by a Curtiss OX-5, 90 horsepower (hp) engine it could reach 75 MPH and stay aloft for 2.5 hours.

111

DeHavilland DH-4 (1917)

Top Speed: 128 MPH
Cruise: 90 MPH
Ceiling: 19,600 ft
Engine: Liberty V-12, 421 hp

The DH-4 was the only US-built airplane to get into combat during World War I. Following World War I, the DH-4 continued in use with the Army for a decade. More than 1,500 were rebuilt to increase their strength and some were modified to carry the airmail in the 1920s.

SPAD XIII (1918)

Top Speed: 138 MPH
Cruise: 124 MPH
Ceiling: 22,300 ft
Engine: Hispano-Suiza 8BEc,
**   235 hp**

The Spad XIII was the best French fighter plane of World War I and the most common fighter-type flown by the American Expeditionary Force (AEF). Fifteen of the 16 AEF pursuit squadrons flew Spad XIIIs by the time the war ended. The main advantage of the Spad XIII over previous Spads was that it had two machine guns instead of one. Capt Eddie Rickenbacker, commander of the 94th Aero Squadron and America's "Ace of Aces," recorded most of his victories in the Spad XIII.

Pursuit/Fighter Aircraft ━━━━━

Curtiss P-6E Hawk (1929)

Top Speed: 204 MPH
Cruise: 167 MPH
Ceiling: 24,900 ft
Engine: Curtiss V-1570-23, 600 hp

    The Curtiss Hawk was a composite of the merits of the best pursuit aircraft of its day. The turbosupercharged V-12 engine served as the predecessor for the V-12 fighter engines that powered World War II aircraft. Lt Col Henry H. "Hap" Arnold designated the new aircraft the P-6E. The Hawk was easy to fly but was usually outmaneuvered in mock dogfights by lighter Army aircraft. This pursuit aircraft was the Army's last major biplane purchase.

Boeing P-26 Pea Shooter (1933)

Top Speed: 234 MPH
Cruise: 199 MPH
Ceiling: 28,300 ft
Engine: P&W R-1340-27,
        600 hp

    The Pea Shooter was the first all-metal monoplane in Air Corps pursuit-squadron service. It was the last Army fighter with an open cockpit, fixed landing gear, and external wires to brace the wings. After the first production delivery of the aircraft, flaps were added to reduce landing speed and the headrest was raised to protect the pilot because the plane often over-turned. The Pea Shooter was obsolete before the United States entered World War II, but Filipino pilots took it into combat against Japanese bombers in 1941.

Bell P-39 Airacobra (1940)

Top Speed: 375 MPH
Cruise: 308 MPH
Ceiling: 33,200 ft
Engine: Allison V-1710-35;
1,150 hp
Range: 450 miles normal

The Airacobra was originally designed to attack enemy aircraft. The first design included a turbosupercharger that was omitted in the P-39C to make the aircraft less expensive and handier at low altitudes. The trade-off was poor high-altitude performance. Most P-39s were lend-leased to the Soviet Union during World War II. Alexander Pokryshkin used a P-39 to score 47 aerial victories making him the second greatest Soviet ace and the most successful ace using an American fighter. The 37 millimeter (mm) cannon in the propeller hub made the P-39 a devastating ground attack aircraft as well.

Lockheed P-38 Lightning (1941)

Top Speed: 414 MPH
Cruise: 290 MPH
Ceiling: 44,000 ft
Engine: (2) Allison V-1710-89;
1,425 hp
Range: 450 miles loaded

The Lightning, designed for pursuit intercept, was the first fighter fast enough to encounter compressibility and consequently suffered buffeting when diving too steeply. The problem was solved using wing fillets. Two drop tanks gave the Lightning a combat radius of 795 miles making it the first fighter useful for long-range escort. The P-38's main limitation was the awkward maneuverability inherent to the aircraft's size. With proper tactics emphasizing its superior speed and ceiling, these limitations were overcome and the Lightning enjoyed great success in the Pacific theater, where range was important.

Curtiss P-40 Warhawk (1939)

Top Speed: 354 MPH
Cruise: 277 MPH
Ceiling: 29,000 ft
Engine: Allison V-1710-39;
1,150 hp
Range: 700 miles

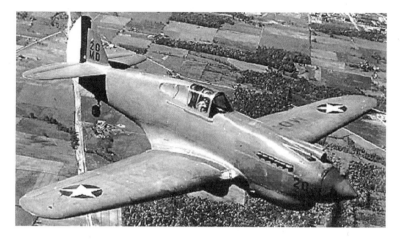

The P-40 was originally dubbed the Kittyhawk but was renamed Warhawk in January 1942. Gen Claire L. Chennault's AVG, "Flying Tigers" flew the P-40 with great success against the Japanese largely because of the superior tactics he devised. The Allison engine had poor high-altitude performance and the P-40 had overall maneuverability limitations. Its strengths were speed, toughness, and firepower. Lt Boyd Wagner shot down eight enemy planes with a P-40E to become World War II's first Army Air Forces (AAF) ace. The P-40 helped the British defeat Erwin Rommel in North Africa.

Republic P-47 Thunderbolt (1942)

Top Speed: 428 MPH
Cruise: 350 MPH
Ceiling: 42,000 ft
Engine: P&W R-2800-59;
2,100 hp
Range: 475 miles with 500 lb
bombs

The Thunderbolt was affectionately called the "Jug." The huge radial engine with bigger turbo was built to speed the aircraft to 15,000 feet under five minutes. It, like all prewar fighters, was designed as an interceptor. The P-47Ds were delivered to Ninth Air Force to support the D day landing in France. The bubble canopy seen in this P-47 was so superior to the old "razorback" shape that every model after the P-47D block 25 was fitted with it. The heavily armed Thunderbolt was outstanding in its role as a low-altitude fighter-bomber in the last year of the European war.

North American P-51 Mustang (1942)

Top Speed: 437 MPH
Cruise: 362 MPH
Ceiling: 41,900 ft
Engine: Packard V-1650-7; 1,490 hp
Range: 950 miles

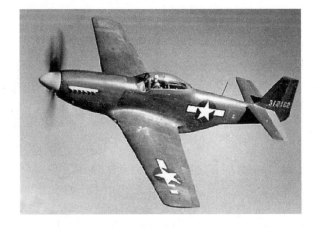

The Mustang was the most effective American air superiority fighter of World War II. At low altitudes, the early Mustangs were superior to the Kittyhawk, Airacobra, and Spitfire in speed and maneuverability. However, the 1,150 hp Allison engine that was initially used had severe limitations at altitude. In the summer of 1942, the Mustang was fitted with the Packard Merlin V-1650-3 engine and a two-stage supercharger. The refitted Mustangs then became the fastest fighters in combat but more importantly the superior design allowed more fuel to be carried. This gave the Mustang the range to escort bombers to any target in Germany.

Bell P-59 Airacomet (1944)

Top Speed: 413 MPH
Cruise: 375 MPH
Ceiling: 46,200 ft
Engine: (2) GE J31-GE-5;
**   2,000 lbs thrust**
Range: 525 miles

The Airacomet was the United States' first jet fighter. Limitations in the jet's performance and tendency toward "snaking" made it a bad gun platform and confined its use to orientation experience with jet flying.

Northrop P-61 Black Widow (1943)

Top Speed: 430 MPH
Cruise: 307 MPH
Ceiling: 41,000 ft
Engine: (2) P&W R-2800-65;
**   2,000 hp**
Range: 415 miles

    The Black Widow was the first American fighter designed specifically for the purpose of night fighting. The plane was equipped with an SCR-720 radar transmitter that permitted location of targets 10 miles away. The P-61 was a very maneuverable aircraft and scored victories against the Japanese in 1944. The three-man crew was composed of pilot, gunner, and radar operator.

Lockheed F-80 Shooting Star (1945)

Top Speed: 580 MPH
Cruise: 439 MPH
Ceiling: 42,750 ft
Engine: Allison J33-A-11;
**   4,000 lbs thrust**
Range: 1,380 miles

    The Shooting Star jet fighter, widely used as the T-33 jet trainer, was the most successful jet produced during World War II. The Shooting Star could match the performance of the ME262. However, the P-80 was delivered in April of 1945 and with the war all but won, it did not see action. The P (pursuit) designation was replaced by F (fighter) in June of 1948. On 8 November 1950, an F-80C destroyed a MiG-15 in air history's first all-jet air battle. Still, mostly unable to compete with the communist fighters, it gave way to the F-86.

Republic F-84 Thunderjet/Thunderstreak (1947)

Top Speed: 622 MPH
Cruise: 483 MPH
Ceiling: 42,100 ft
Engine: Allison J35-A-29;
      5,600 lbs thrust
Range: 2,000 miles

The F-84 was the first new fighter of the postwar period. The early straight-wing versions of the F-84 were dubbed "Thunderjets." Subsequent modifications improved the jet's power bringing it close to the airframe's speed limit. The F-84F was designed with swept-wings and called the "Thunderstreak." They were equipped with a low-altitude bombing system (LABS) to deliver tactical nuclear weapons. The F-84 saw extensive action in Korea as a fighter-bomber, but, it too, could not compete with MiG-15s in air-to-air combat.

North American F-86 Sabre (1948)

Top Speed: 695 MPH
Cruise: 486 MPH
Ceiling: 48,000 ft
Engine: GE J47-GE-27;
      5,910 lbs thrust
Range: 458 miles combat
      radius

The Sabre was the first American aircraft to be designed with swept wings. This modification delays the onset of compressibility troubles inherent with flying at sonic speeds. Although the F-86 was inferior in climb and ceiling to the lighter Soviet type fighters, with better pilots it proved superior in air-to-air combat. In Korea, Sabres downed 792 MiGs while losing only 78.

North American F-100 Super Sabre (1954)

Top Speed: 924 MPH
Cruise: 593 MPH
Ceiling: 38,700 ft
Engine: P&W J57-P-21;
       10,200 lbs thrust
Range: 572 miles combat
       radius

The Super Sabre was the first supersonic aircraft in production anywhere in the world. The F-100 was the first in the Air Force's "Century" series of fighters. The F-100 created maintenance and cost problems for the Air Force. Titanium was used in construction, raising the cost to over $1 million per aircraft. Old by the 1970s, the F-100 was used in Vietnam as high-speed, forward air controllers (FAC). The Fast FACs and other F-100s were used operationally until 1972.

McDonnell F-101 Voodoo (1955)

Top Speed: 1,005 MPH
Cruise: 550 MPH
Ceiling: 38,900 ft
Engine: (2) P&W J57-P-13;
       10,200 lbs thrust
Range: 677 miles combat
       radius

The Voodoo was the first 1,000-MPH fighter in production. The F-101 was intended to serve as Strategic Air Command's (SAC) fighter-escort but was transferred to Tactical Air Command (TAC) as a low-altitude fighter-bomber to carry Mk 7 nuclear bombs.

Convair F-102 Delta Dagger (1956)

Top Speed: 780 MPH
Cruise: 605 MPH
Ceiling: 53,400 ft
Engine: P&W J57-P-23; 10,200 lbs
  thrust
Range: 386 miles combat radius

  The F-102 was the product of the Air Force's intention to create an air defense fighter interceptor. It used guided missiles instead of guns or unguided rockets. The delta wing design originally had limitations, particularly a drag hump at sonic speed. A design fix saved the program and the Delta Dagger served Air Defense Command (ADC) squadrons through 1973.

Lockheed F-104 Starfighter (1958)

Top Speed: 1,324 MPH
Cruise: 584 MPH
Ceiling: 58,000 ft
Engine: GE J79-GE-7; 15,800 lbs
  thrust
Range: 352 miles combat radius

  The Starfighter was the first Air Force fighter with Mach 2 speed. The F-104 had the smallest and thinnest wings ever used on an American fighter. In 1958, it became the first aircraft to set both the speed and altitude record. However, a small weapons load and limited endurance made them unsuitable for the Vietnam War. Though the F-104 was passed onto the Air National Guard in 1967, it was purchased by 14 foreign air forces and served into the 1980s.

Convair F-106 Delta Dart (1959)

Top Speed: 1,328 MPH
Cruise: 594 MPH
Ceiling: 52,700 ft
Engine: P&W J75-P-17;
  16,100 lbs thrust
Range: 364 miles combat radius

  The Delta Dart was a further refinement of the F-102B design. As intercontinental ballistic missiles (ICBM) replaced bombers as the principle nuclear threat, the need for interceptors was reduced. The F-106 remained in ADC in diminishing numbers through the 1970s.

Republic F-105 Thunderchief (1959)

Top Speed: 1,372 MPH
Cruise: 778 MPH
Ceiling: 48,500 ft
Engine: P&W J75-P-19W;
      6,100 lbs thrust
Range: 778 miles combat radius

    The nuclear-carrying "Thud" contained the first internal bomb bay used on a fighter. The F-105 made the first supersonic bomb drops. As emphasis shifted from nuclear warfare to conventional bombs, the Thud's internal bomb bay was closed and a centerline pylon was added. All-weather capabilities and speed combined with a toss-bomb computer made it the premier fighter-bomber of the Vietnam War. The F-105D delivered 75 percent of the bomb tonnage of Rolling Thunder.

McDonnell Douglas F-4 Phantom II (1962)

Top Speed: 1,400 MPH
Cruise: 590 MPH
Ceiling: 55,850 ft
Engine: (2) General Electric
      J79-GE-15; 17,000 lbs
      thrust
Range: 1,750 miles

    The F-4 Phantom opened the era of the Mach 2 missile-launching fighter. In 1960, the F-4 made a number of record-breaking flights: a world-speed record (1,606 MPH or Mach 2.57), a climb to 49,212 feet in 114 seconds, and a zoom climb to 98,556 feet. The first operational version was the F-4B, a Navy all-weather fighter. The Air Force purchased the F-4C in 1962. The original F-4Cs that went to Vietnam in 1965 did not have guns. An alert enemy pilot making a sharp turn could evade the Phantom's missiles. A 20-mm external gun pod was added in 1967. F-4Ds with laser-guided 3,000-lb bombs finally destroyed the Thanh Hoa Bridge. Of 137 United States Air Force (USAF) aerial victories in Vietnam, Phantoms were credited with 107.5. "Wild Weasel" F-4G aircraft provided lethal air defense suppression for deep strikes on Iraq during Operation Desert Storm.

General Dynamics F-111 Aardvark (1967)

Top Speed: 1,453 MPH
Cruise: 498 MPH
Ceiling: 57,900 ft
Engine: (2) P&W TF30-P-100;
**   25,000 lbs thrust**
Range: Combat radius, 700
**   miles with 1,200 lbs**
**   bombs**

The F-111 program was extremely ambitious and controversial. The goal was to produce a common fighter-bomber for both the Air Force and the Navy. The most unusual feature of the design was a "swing-wing" that could be swept forward for take-off, landing, and low speed operations and swept back for high speed operations. The attempt to make the F-111 all things to all people resulted in a plane that was too heavy for carrier operations so the Navy developed a pure air-superiority swing-wing fighter (the F-14) instead. The F-111 was not an effective air-superiority fighter but its high-speed, long-range, heavy bomb load, and all-weather navigation and attack capability made it an effective bomber during and after the Vietnam War. The electronic warfare version of the F-111 played a major role in Operation Desert Storm.

McDonnell Douglas F-15 Eagle (1976)

Top Speed: 1,875 MPH
Cruise: 570 MPH
Ceiling: 65,000 ft
Engine: (2) P&W F100-PW-100;
**   25,000 lbs thrust**
Range: Ferry 3,450 miles

The F-15 is the world's premier air-superiority fighter achieving 32 aerial victories during Operation Desert Storm with no air-to-air losses. Its advanced pulse-Doppler radar work with the AIM-7 and AIM-120 air-to-air missiles to destroy enemy aircraft beyond visual range. Within visual range, the F-15's outstanding acceleration, speed, and maneuverability (due to its high thrust-weight ratio and large wings) enable it to outmaneuver its foes so they can be destroyed by the F-15's AIM-9 missiles and the M-61 Vulcan cannon. The F-15E is a two-seat strike version of the F-15, capable of devastating ground attacks deep in enemy territory.

Lockheed Martin F-16 Fighting Falcon (1979)

Top Speed: 1,500 MPH
Ceiling: >50,000 ft
Engine: General Electric
F110-GE-129
Range: Ferry >2,000 miles

The F-16 was designed as a light, relatively low cost, single-seat multirole fighter. It has been so successful that the USAF has more F-16s than any other type of aircraft and the F-16 is the backbone of many allied air forces. F-16s proved their worth in combat with the Israeli air force and during Operation Desert Storm where F-16s flew more sorties than any other aircraft type. Continuing improvements to the F-16 will enable it to serve as a workhorse of the Air Force for several more decades.

Lockheed F-117 Nighthawk (1983)

Top Speed: NA
Cruise: 580 MPH
Ceiling: NA
Engine: (2) General Electric
F404-GE-F1D2; 10,500
lbs thrust
Range: Combat radius 750 miles

The F-117 represents a revolution in aircraft design. It was developed under conditions of extreme secrecy at the famous Lockheed "Skunk Works." It is the first production aircraft designed specifically to take advantage of breakthroughs in low-observable (stealth) technology. The F-117's unique shape makes it nearly invisible to radar while its engines give off a minimal infrared signature and flying at night prevents visual detection. During Desert Storm the F-117 few over twelve hundred strike missions against the most heavily defended Iraqi targets without losing a single aircraft.

123

Boeing and Lockheed Martin F-22 Raptor

Top Speed: Mach 2
Cruise: Mach 1.5
Ceiling: >65,000 ft
Engine: (2) P&W F119-PW-100;
        35,000 lbs thrust

The F-22 is the next generation air-superiority fighter, designed to defeat any adversary's aircraft and dominate the skies of the early twenty-first century. It possesses a unique combination of stealth, maneuverability, super-cruise, and integrated avionics: truly leap-ahead technologies that make it vastly more capable than any previous aircraft. Stealth, or more accurately "low observable," technology makes the F-22 very difficult to identify by radar or other means. The F-22's unique engines provide extraordinary maneuverability through "thrust vectoring": not all the engine thrust is always going straight ahead, but is instead redirected, to enhance the maneuvers commanded by the pilot. These engines also enable "super-cruise": supersonic flight maintained for long periods of time because the engines achieve it without afterburners. Integrated avionics enable each F-22 to receive automated information (such as the radar picture) from other F-22s in its formation and thus dramatically enhances teamwork within the formation. Though originally intended as a pure air-superiority fighter, the F-22's inherent ground attack capability is being developed through adding the joint direct attack munition to its armament.

Martin GMB/MB-1 (1918)

Top Speed: 105 MPH
Cruise: 100 MPH
Ceiling: 12,250 ft
Engine: (2) Liberty 12A; 400 hp
Range: 390 miles with 1,040 lbs bombs

The MB-1 and the improvements embodied in the MB-2, offered the Army a platform to demonstrate the capability of aerial bombing. In 1921, William "Billy" Mitchell used MB-2s to sink the ex-German battle cruiser *Ostfriesland*.

Boeing B-9 (1932)

Top Speed: 188 MPH
Cruise: 165 MPH
Ceiling: 22,500 ft
Engine: (2) P&W R-1860-11;
      600 hp
Range: 540 miles with 2,260 lbs
      bombs

With the world-class Pratt & Whitney radial engines, the B-9 could fly 50 percent faster than contemporary bombers. This made bomber speed comparable to pursuit aircraft speed. The B-9 monoplane design became the parent of World War II bombers.

Martin B-10 (1935)

Top Speed: 213 MPH
Cruise: 193 MPH
Ceiling: 24,200 ft
Engine: (2) Wright R-1820-33; 775 hp
Range: 1,240 miles with 2,260 lbs
      bombs

The B-10 was the first all-metal monoplane bomber to be produced in quantity. It also introduced such innovations as a monocoque fuselage, variable-pitch propellers, retractable landing gear, enclosed cockpits, and a rotating gun turret. In 1934, Colonel Arnold led a mass flight of B-10 crews 18,000 miles from Washington, D.C., to Alaska, proving the capability for strategic bombing.

Boeing B-17 Flying Fortress (1939)

Top Speed: 287 MPH
Cruise: 182 MPH
Ceiling: 35,600 ft
Engine: (4) Wright R-1820-97;
        1,200 hp
Range: 2,000 miles with 6,000
        lbs bombs

The Flying Fortress, with a 10-12 man crew, was the first truly modern heavy bomber and one of the most recognized airplanes of the World War II era. It was the Army's first production four-engine bomber. The YB-17 made its first flight in 1935 but the B-17C ordered in 1939 was the first into combat in World War II. Total B-17 production was 12,276 with peak AAF inventory reaching 4,574 in August 1944.

Consolidated B-24 Liberator (1941)

Top Speed: 303 MPH
Cruise: 200 MPH
Ceiling: 32,000 ft
Engine: (4) P&W R-1830-65;
        1,200 hp
Range: 2,300 miles with 5,000
        lbs bombs

More Liberators were built than any other American airplane in history (18,482), with peak inventory reaching 6,043 in September 1944. The B-24 was used in every theater in World War II, and it had greater range and could carry a much larger bomb load than the B-17. It had a crew of 10. The B-24's most famous missions were the raids on Ploesti oil fields in Romania.

North American B-25 Mitchell (1940)

Top Speed: 272 MPH
Cruise: 230 MPH
Ceiling: 24,200 ft
Engine: (2) Wright R-2600-29; 1,700 hp
Range: 1,300 miles with 3,000 lbs bombs

Named in honor of US airpower proponent Brig Gen William "Billy" Mitchell, the B-25 served in every theater of World War II and was made in larger quantities than any other American twin-engine combat airplane. On 18 April 1942, Lt Col James H. "Jimmy" Doolittle's B-25 crews took off from an aircraft carrier and bombed Tokyo and other targets, the first time US aircraft had bombed Japan.

Martin B-26 Marauder (1941)

Top Speed: 282 MPH
Cruise: 214 MPH
Ceiling: 21, 700 ft
Engine: (2) P&W R-2800-43;
**   2,000 hp**
Range: 1,150 miles with 3,000
**   lbs bombs**

The B-26 was ordered off the drawing board (no prototypes were built) at the same time as the B-25. With a troubled development history, it was called the "Flying Prostitute"— with its high wing loading (51 lbs per square feet) and small wings, it was said to have had no visible means of support. Less Marauders were built than B-25s because the B-26 proved to be more expensive to build, maintain, and had higher accident rates. The Marauder had a great war record, bringing back a higher percentage of crews than B-17s or B-24s.

Boeing B-29 Superfortress (1944)

Top Speed: 399 MPH
Cruise: 253 MPH
Ceiling: 36,150 ft
Engine: (4) Wright R-3350-57;
**   2,200 hp**
Range: combat radius, 1,931
**   miles**

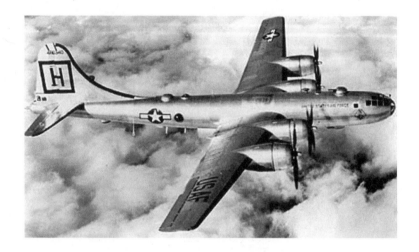

The Superfortress was the best bomber of World War II and the first bomber to include pressurized crew compartments and remote controlled gun turrets. On the night of 9 March 1945, B-29s conducted the most devastating air raid in history, destroying almost 16 square miles of Tokyo. In August 1945, specially modified B-29s dropped two atomic bombs on Japan, ending World War II. During the Korean War B-29s also did extensive conventional bombing. B-50s (B-29s with improved engines) were modified as the first aerial tankers used in large numbers and served as weather aircraft into the 1960s.

Convair B-36 Peacemaker (1948)

Top Speed: 418 MPH
Cruise: 227 MPH
Ceiling: 43,600 ft
Engine: (6) P&W R-4360-53;
**   3,800 hp (4) General**
**   Electric J47-GE-19;**
**   5,200 lbs thrust**
Range: 3,990 miles combat
**   radius with 10,000 lbs**
**   bombs**

When it was first flown, the B-36 was the largest aircraft in the world. With its intercontinental range and ability to carry nuclear weapons (primarily the Mk 17 hydrogen bomb), the B-36 served as the United States' airborne nuclear deterrent through the 1950s. The Peacemaker was politically controversial, and caused the "Revolt of the Admirals" in 1949 when it was acquired at the expense of a Navy Super Cruiser.

Boeing B-47 Stratojet (1951)

Top Speed: 607 MPH
Cruise: 498 MPH
Ceiling: 40,500 ft
Engine: (6) General Electric J47-GE-25; 7,200 lbs thrust
Range: 2,013 miles combat radius with 10,845 lbs bombs

The B-47 was the world's first swept-wing bomber and the first to use a bicycle arrangement for the landing gear, which was necessary because of the thin wing. At the time, the design was so advanced that some writers called it "futuristic." Over 2,000 entered service with SAC and the B-47 was the nation's primary nuclear deterrent until it was replaced by the B-52 and intercontinental ballistic missiles.

Convair B-58 Hustler (1960)

Top Speed: 1,321 MPH
Cruise: 610 MPH
Ceiling: 64,800 ft
Engine: (4) General Electric J79-GE-5B; 15,300 lbs thrust
Range: 1,750 miles combat radius

The B-58 was the first supersonic bomber put into production and the first bomber to reach Mach 2. The Hustler used a "pod" instead of internal weapons stowage and made use of stainless-steel honeycomb construction for lower weight and greater strength. Crew members had individual escape capsules rather than individual ejection seats. The B-58 used sophisticated inertial and star-tracking navigation methods. On 15 October 1959, one of the first B-58s built was flown 1,680 miles in 80 minutes with one refueling, maintaining a speed of more than Mach 2 for more than an hour.

Boeing B-52 Stratofortress (1955)

Top Speed: 650 MPH
Cruise: 530 MPH
Ceiling: 50,000 ft
Engine: (8) P&W J-57-P-43WB;
**   13,750 lbs thrust**
Range: 3,500 miles combat
**   radius**

The enormous range and bomb load of the B-52 has made it one of the US Air Force's most successful aircraft. Though originally designed as a high-altitude nuclear bomber, the B-52 has also proven effective in low altitude and conventional bombing operations. It is the longest serving combat aircraft in the history of the USAF and it will continue to be a major part of the US bomber force well into the twenty-first century. In addition to serving as a cornerstone of the US nuclear deterrent, B-52s conducted devastating bombing missions during both the Vietnam War and Operation Desert Storm.

Rockwell B-1 Lancer (1985)

Top Speed: >900 MPH
Cruise: high subsonic
Ceiling: >50,000 ft
Engines: (4) GE F-101-GE-102;
**   >30,000 lbs thrust**
Range: >7,000 miles

The B-1 was the first US Air Force bomber to incorporate low-observable stealth technology and was designed for low-altitude, deep-penetration nuclear attacks on the Soviet Union. It has a swing-wing that gives it outstanding performance at all speeds and altitudes. With the end of the Soviet threat, nonnuclear bombing capability has been added to the B-1.

Northrop Grumman B-2 Spirit (1999)

Top Speed: High Subsonic
Ceiling: 50,000 ft
Engine: (4) GE
F-118-GE-100
17,300 lbs
Range: 6,000 miles

The B-2 is the first operational combat aircraft ever built according to the "flying wing" concept (i.e., without a tail or fuselage). Its unique shape and the exotic materials used in its construction make it extremely difficult to detect with radar. This extraordinary level of stealth is achieved in spite of the fact that the B-2 is a very large aircraft capable of delivering more than 32,000 lbs of bombs on enemy targets with great precision. Though originally designed to attack targets within the Soviet Union with nuclear weapons, the B-2 made its combat debut during Operation Allied Force when B-2's delivered 2,000 lb joint direct attack munitions (JDAM) on targets in Serbia. The B-2's had to fly out of their home base—Whiteman AFB, Missouri—so each mission took over 30 hours. In spite of the low sortie rate caused by these long missions and the small number of B-2's operational in 1999, the ability of each B-2 to deliver 16 JDAMs with extraordinary accuracy in all weather on the most heavily defended targets made it the preferred system for attacking many targets in Serbia.

Attack Aircraft

Northrop A-17 Nomad (1936)

Top Speed: 206 MPH
Cruise: 170 MPH
Ceiling: 20,700 ft
Engine: P&W R-1535-11;
        750 hp
Range: 650 miles with 654 lbs
        bombs

Northrop developed the A-17 in the early 1930s to fill an Army Air Corps (AAC) requirement for a ground-attack bomber. The aircraft would provide ground support for infantry, attacking at tree-top level with its four forward-firing .30-caliber machine guns and fragmentation or chemical bombs carried internally.

Douglas A-20 Havoc (1940)

Top Speed: 317 MPH
Cruise: 230 MPH
Ceiling: 25,000 ft
Engine: (2) Wright R-2600-23;
        1,600 hp
Range: 1,025 miles

The A-20 was designed to meet an AAC attack specification in 1938 but it took British and French orders to start production. Begun as a company-funded venture, the Havoc eventually became the most produced AAF attack aircraft. It also helped start the trend for nose wheels in American combat aircraft.

Douglas A-26 Invader (1941)

Top Speed: 355 MPH
Cruise: 284 MPH
Ceiling: 22,100 ft
Engine: (2) P&W R-2800-27;
       2,000 hp
Range: 1,400 miles with
       4,000 lbs bombs

    An updated A-20, it was used for level bombing, ground strafing, and rocket attacks. When production halted after VJ day, 2,502 Invaders had been built. The A-26 was redesignated the B-26 in 1948 (forever confusing it with the Martin B-26). During the Korean War, the airplane entered combat once again, this time as a night intruder to harass North Korean supply lines. Early in the Vietnam War, the Invader went into action for the third time. In 1966, the B-26K was redesignated the A-26A.

Douglas A-1 Skyraider (1963)

Top Speed: 325 MPH
Cruise: 240 MPH
Ceiling: 26,200 ft
Engine: Wright R-3350; 2,700 hp
Range: 1,500 miles

    This rugged dump truck of an attack aircraft was originally developed for the Navy and entered Navy service in the 1940s. The Air Force used A-1s to attack ground targets in Vietnam, Cambodia, and Laos (where it worked closely with the Central Intelligence Agency's [CIA] secret operations). It was nicknamed "Spad," because pilots considered it a throwback airplane in the jet-age Air Force. The Skyraider was also used to cover rescue operations, where the type picked up a second nickname, "Sandy," its radio call sign.

Vought A-7 Corsair II (1968)

Top Speed: 663 MPH
Cruise: 545 MPH
Ceiling: 33,500 ft
Engine: Allison TF41;
14,250 lbs thrust
Range: 1,330 miles with
6,560 lbs bombs

The A-7 was originally designed for the US Navy which began purchasing them in 1965. The A-7 was the USAF's first jet attack aircraft and performed well in the later stages of the Vietnam War. The A-7 was the Air Force's primary ground attack aircraft throughout the 1970s.

Fairchild A-10 Thunderbolt II (1976)

Top Speed: 450 MPH
Cruise: 335 MPH
Ceiling: 44,200 ft
Engine: (2) GE TF34-GE-100;
9,000 lbs thrust
Range: 288 miles combat
radius with 9,540 lbs
bombs and 2 hours on
station

The A-10 was designed to defeat a Soviet invasion of Western Europe by killing huge numbers of Soviet tanks. To do this it was equipped with one of the most powerful cannons ever mounted in an aircraft in addition to bombs and ground attack missiles. The A-10 relies on its exceptional maneuverability, heavy armor, and redundant systems, rather than speed, for its survivability. Though the Soviets never invaded, the A-10 demonstrated its effectiveness against Iraqi forces during Operation Desert Storm. Though its lack of night and all-weather capability limit its effectiveness, the A-10's heavy firepower, long station time, and ease of maintenance have made it popular on peace enforcement missions in places like Bosnia.

Tankers

Boeing KC-97 Stratofreighter (1950)

Top Speed: 400 MPH
Cruise: 230 MPH
Ceiling: 30,000 ft
Engine: (4) P&W R-4360;
   3,500 hp (2) GE-J47s,
   5,970 lbs thrust
Range: 2,300 miles

The KC-97, a tanker version of the C-97, was introduced in 1950 using the "flying boom" refueling system. This was an improvement from the probe-and-drogue technique tested earlier. The value of air refueling was demonstrated in 1952 by nonstop deployment of the 31st Fighter-Escort Wing from Turner AFB, Georgia, to Japan. Jet squadrons could now fly to any point in the world refueling from flying tankers at prearranged rendezvous points. The jet engines were added in 1964 to give the Stratofreighter enough speed to refuel faster jet aircraft.

Boeing KB-50 Superfortress (1956)

Top Speed: 445 MPH
Cruise: 410 MPH
Ceiling: 40,000 ft
Engine: (4) P&W R-4360;
   3,500 hp (2) GE-J47s;
   5,910 lbs thrust
Range: 2,500 miles

KB-50s were B-50 bombers modified as aerial tankers. The alterations involved the removal of armament and the installation of additional fuel tanks and probe-and-drogue equipment to permit the aerial refueling by hose of up to three fighter-type aircraft at one time. As the performance of operational jet fighters increased, a J47 jet engine was installed under each wing on some KB-50s to boost speed while refueling and to increase altitude capability.

Boeing KC-135 Stratotanker (1957)

Top Speed: 610 MPH
Cruise: 530 MPH
Ceiling: 50,000 ft
Engine: (4) CFM-56;
      21,634 lbs thrust
Range: 1,500 miles with
      150,000 lbs fuel

The KC-135 derives from the same prototype as the Boeing 707 commercial airplane and uses the flying boom refueling system. It began replacing the KC-97s of the Strategic Air Command in the late 1950s and later replaced Tactical Air Command's KB-50s. KC-135s played a central role in Air Force combat operations in Vietnam, Panama (1989), and Operations Desert Shield and Desert Storm. There are more than 600 KC-135s in the Air Force inventory and the new CFM-56 engines will enable them to remain the backbone of the refueling force well into the twenty-first century. The EC-135 provides airborne command and control for US nuclear forces in the event ground command and control is knocked out.

McDonnell Douglas KC-10 Extender (1981)

Top Speed: 619 MPH
Cruise: 564 MPH
Ceiling: 42,000 ft
Engine: (3) GE CF-6-
      50C2; 52,500
      lbs thrust
Range: 4,400 miles
      with cargo

The KC-10 is a modified civilian DC-10. It is capable of aerial refueling using both flying-boom and hose-and-drogue refueling systems. When used strictly as a refueling aircraft, the KC-10 can carry almost twice as much fuel as a KC-135. The KC-10 can also be configured to transport personnel and equipment while conducting its primary aerial refueling mission. During Operations Desert Shield and Desert Storm, KC-10s refueled aircraft from all branches of the US armed forces and many coalition forces.

Transport and Special Electronic Aircraft

Curtiss-Wright C-30 Condor (1933)

Top Speed: 181 MPH
Ceiling: 22,800 ft
Engine: (2) Wright R-1820-12;
          700 hp
Range: 750 miles

The older Army transports, like the C-30, were procured in small numbers and essentially standard commercial aircraft modified to meet military requirements. In some cases larger cargo doors were fitted; in others unusual engines were replaced by standard Army models.

Douglas C-47 Skytrain "Gooney Bird" (1941)

Top Speed: 232 MPH
Cruise: 175 MPH
Ceiling: 24,450 ft
Engine: (2) P&W R-1830;
          1,200 hp
Range: 1,513

Few aircraft are as well known or were so widely used for so long as the C-47 or "Gooney Bird" as it was affectionately nicknamed. The first C-47s were ordered in 1940 and by the end of World War II, 9,348 had been procured for AAF use. After World War II, many C-47s remained in USAF service, participating in the Berlin Airlift and other peacetime activities. During the Korean War, C-47s hauled supplies, dropped paratroops, evacuated wounded, and dropped flares for night bombing attacks. In Vietnam, the C-47 served again as a transport, but it was also used in a variety of other ways which included flying ground attack (gunship), reconnaissance, and psychological warfare missions.

Curtiss C-46 Commando (1942)

Top Speed: 245 MPH
Cruise: 175 MPH
Ceiling: 27,600 ft
Engine: (2) P&W R-2800;
      2,000 hp
Range: 1,200 miles

    The C-46 gained its greatest fame during World War II transporting war materials over the "Hump" from India to China after the Japanese had closed the Burma Road. The Commando carried more cargo than the famous C-47 and offered better performance at higher altitudes, but under these difficult flying conditions, C-46s required extensive maintenance and had a relatively high loss rate. During the Korean War C-46s saw additional service.

Douglas C-54 Skymaster (1942)

Top Speed: 300 MPH
Cruise: 245 MPH
Ceiling: 30,000 ft
Engine: (4) P&W R-2000;
      1,450 hp
Range: 3,900 miles

    This long-range heavy transport gained its greatest fame in World War II, the Berlin Airlift, and the Korean War. Originally developed for the airlines, the first batch of what would have been DC-4s was commandeered off the assembly line by the Army Air Forces in 1942 and redesignated C-54. The first presidential aircraft was the lone VC-54C. President Harry S. Truman signed the National Security Act of 1947, creating an independent Air Force, on board this aircraft on 12 July 1947.

Fairchild C-82/119 Flying Boxcar (1945)

Top Speed: 290 MPH
Cruise: 200 MPH
Ceiling: 30,000 ft
Engine: (2) Wright R-3350;
        3,500 hp
Range: 2,000 miles

The C-82 appeared too late to participate in World War II and the C-119 was a major redesign of the C-82. The C-119 had the same major design feature as the C-82—a rear-loading, all-through cargo hold—but featured more powerful engines and a relocated flight deck. In the 1950s, C-119s were used to ferry supplies to the Arctic for construction of the distant early warning (DEW) line radar sites. In the late 1960s, the Air Force selected the C-119 to replace the AC-47 Spooky in the gunship role.

Douglas C-124 Globemaster "Old Shaky" (1950)

Top Speed: 320 MPH
Cruise: 200 MPH
Ceiling: 34,000 ft
Engine: (4) P&W R-4360;
        3,800 hp
Range: 2,175 miles

The Globemaster performed yeoman service through two wars and nearly 25 years. The C-124 was a major redesign of the C-74 that was developed at the end of World War II. It used the same wings, tail, and engines as the C-74 but had a deeper fuselage that featured clamshell doors in the nose to permit vehicles to drive on and off under their own power. The C-124 retained the C-74's electrically operated elevator in the rear of the aircraft for loading of bulk cargo. It went on to provide a much-needed airlift capability in the Korean War, as it was the only aircraft that could carry many of the Army's vehicles.

Lockheed C-69/121 Constellation (1943)

Top Speed: 290 MPH
Cruise: 240 MPH
Ceiling: 18,000 ft
Engine: (4) Wright R-3350;
      3,400 hp
Range: 4,000 miles

    Howard Hughes was one of the driving forces behind the design of the Lockheed Constellation. A number of aircraft intended for Hughes' Transcontinental and Western Airlines (and for Pan American) were requisitioned off the production line by the United States AAF after the attack on Pearl Harbor and designated C-69s. It was the heaviest and the fastest transport built to date for the Army Air Forces. The improved Constellation (the C-121) entered service after World War II and most of them were built as EC-121s and used for electronic reconnaissance and airborne early warning. In the mid-1960s, the Air Force sent the first Warning Stars to Vietnam to maintain radar surveillance over North Vietnam and then later to warn of MiG attacks.

Fairchild C-123 Provider (1955)

Top Speed: 240 MPH
Cruise: 170 MPH
Ceiling: 28,000 ft
Engine: (2) P&W R-2800;
      2,500 hp (2) GE J-85;
      2,850 lbs thrust
Range: 1,825 miles

    The C-123 was a tactical transport originally designed as a glider, although the design was drawn up with the intention of its eventually being powered. Two C-123Ks were modified to the NC-123K configuration (also referred to as AC-123K) under the Black Spot project. This was designed to give the Air Force a self-contained night attack capability to seek out and destroy targets on the Ho Chi Minh Trail. The C-123s contributed a substantial portion of in-country airlift and resupply in Vietnam and Cambodia.

Lockheed C-130 Hercules (1957)

Top Speed: 380 MPH
Cruise: 340 MPH
Ceilings: 33,000 ft
Engines: (4) Allison T-56-A-15;
**   4,300 hp**
Range: 2,500 miles with
**   25,000 lbs cargo**

The C-130's primary mission is airlift within theater where its ability to operate from rough, dirt airstrips is vital. It is also the primary aircraft for paradropping troops and equipment. C-130s dropped Army troops into combat in Grenada and Panama. The rugged and versatile C-130 has been modified to perform aerial refueling (KC-130) and electronic warfare (EC-130) operations and as the AC-130, it has served as the nation's primary gunship since the early 1970s. The Air Force Special Operations Command relies particularly heavily on C-130 variants. The C-130 has been in production for over four decades, longer than any other aircraft in history.

Lockheed C-141 Starlifter (1964)

Top Speed: 566 MPH
Cruise: 500 MPH
Ceiling: 41,000 ft
Engines: (4) P&W TF33-P-7;
**   20,250 lbs thrust**
Range: 3,000 miles with
**   maximum payload**

The C-141 was Air Mobility Command's first jet aircraft designed to meet military cargo standards. It dropped Army troops and equipment into combat in Panama and Grenada and was critical to airlift operations during Operations Desert Shield/Desert Storm. Between 1979 and 1982 the entire C-141 fleet was extensively modified to add air refueling receptacles and lengthen each fuselage by over 23 feet. It has been the backbone of the US airlift fleet since the mid-1960s and will continue to provide intercontinental airlift until replaced by the C-17.

Lockheed C-5 Galaxy (1970)

Top Speed: 570 MPH
Cruise: 518 MPH
Ceiling: >30,000 ft
Engine: (4) GE TF-39;
41,000 lbs thrust
Range: >2,000 miles with
200,000 lbs of cargo

The C-5 carries outsized cargo intercontinental distances. It is one of the largest aircraft ever built; so large that the Wright brothers' entire first flight could have been conducted inside a C-5's cargo bay. Its upward-lifting "visor" nose, full-width aft doors and ramps, and "kneel down" landing gear speed the loading and unloading of the C-5. The C-5 fleet was critical to US airlift operations during Operations Desert Shield and Desert Storm and there is no scheduled replacement date for the C-5.

Boeing E-3 Sentry AWACS (1977)

Optimal Cruise: 360 MPH
Ceiling: >29,000 ft
Engine: (4) P&W TF33-PW-100A;
21,000 lbs thrust
Endurance: Over 8 hours without
refueling

The E-3 is a modified Boeing 707 with a large rotating radar dome. It serves as an airborne warning and control system (AWACS) providing all-weather surveillance, command, control, and communications to United States and allied aircraft. Since they entered service, E-3s have been among the first aircraft dispatched to world trouble spots and they directed more than 120,000 coalition sorties during Operation Desert Storm.

Northrop Grumman E-8 JSTARS (1991)

Top Speed: 600 MPH
Cruise: 535 MPH
Ceiling: NA
Engines: (4) JT3D-7;
        19,000 lbs thrust
Range: 5,700 miles

The E-8 joint surveillance target attack radar system (JSTARS) is a modified Boeing 707 with a phased-array radar antenna mounted in a 26-foot long canoe-shaped radome under the forward part of the fuselage. The JSTARS radar monitors ground activity deep inside enemy territory and provides target data to both the USAF and US Army. Though it was still under development in 1991, the test bed E-8s were deployed to the Persian Gulf and participated in Operation Desert Storm. Since then they have been very active in other trouble spots.

Boeing C-17 Globemaster III (1993)

Cruise: 500 MPH
Ceiling: 45,000 ft
Engines: (4) P&W F117-PW-100;
        40,440 lbs thrust
Range: 2,800 miles with
        160,000 lbs cargo

The C-17 is gradually replacing the C-141 in the USAF airlift fleet and the C-17 will be the backbone of United States air mobility in the first half of the twenty-first century. The C-17 can carry about twice as much cargo as the C-141 and can operate from shorter, more austere airfields. These features greatly enhance the capabilities of Air Mobility Command.

Training Aircraft

Vultee BT-13 Valiant (1940)

Top Speed: 173 MPH
Engine: P&W R-985-25;
450 hp

The Valiant was the basic trainer most widely used by the USAAF during World War II. Over 10,000 Valiants were built, which provided basic pilot training for the enormous number of pilots need for the war. The BT-13 had a powerful engine and was faster and heavier than other trainers. It required the student pilot to use two-way radio communications with the ground, and to operate both landing flaps and a two-position variable pitch propeller.

North American AT-6 Texan (1940)

Top Speed: 206 MPH
Engine: P&W R-1340-47,
600 hp

The AT-6 was one of the most widely used aircraft in history. Nearly all USAAF pilots in World War II trained in the AT-6 prior to earning their wings. This aircraft exposed students to high-performance aircraft with retractable landing gear. The British version of the AT-6 the "Harvard" was widely used by the British Royal Air Force during the war. During the Korean War, some AT-6s were used to spot enemy ground targets and mark them with smoke rockets for attack by fighter-bombers.

Lockheed T-33 Shooting Star (1948)

Top Speed: 600 MPH
Engine: J33-A-23; 5,200 lbs
thrust

The two-place T-33 was the world's first jet trainer. It was developed from the single-seat F-80 fighter by lengthening the fuselage approximately three feet to accommodate a second cockpit. Lockheed undertook the design of the T-33 with $1 million of its own money. Entering service in 1948, the T-33 was the only Air Force jet trainer until the advent of the Cessna T-37 in 1957.

North American T-28 Trojan (1949)

Top Speed: 285 MPH
Engine: Wright R-1300-1;
800 hp

The winner of the competition to build the first new Air Force trainer after World War II, the T-28A fell short of expectations, and the Air Force eventually added the T-34 to the syllabus to provide ab initio training. The United States used a number of T-28s in the early stages of Operation Farm Gate to train the South Vietnamese in commando-type operations. In 1962, the USAF began to modify nearly three hundred T-28s as fighter-bombers for counterinsurgency warfare, and these were redesignated T-28D Nomads.

Beech T-34 Mentor (1952)

Top speed: 180 MPH
Engine: O-470-13; 225 hp

The T-34A was used by the USAF for primary flight training during the 1950s. The T-34A served as the standard primary trainer until the USAF introduced the Cessna T-37 jet trainer in the late 1950s. As they were replaced by the T-37, many T-34s were turned over to base Aero Clubs. In all, 450 T-34As were produced for the USAF.

Cessna T-37 Tweet (1957)

Speed: 410 MPH
Engine: (2) Continental
        J69-T-25; 1,025
        lbs thrust

The T-37 is one of the most successful training aircraft ever built and has been the primary trainer for several generations of Air Force pilots. The student and instructor sit side by side to facilitate instruction and the twin engines and flying characteristics help prepare students to move on to larger and faster aircraft. Modified T-37s have also served as attack aircraft in counterinsurgency operations.

Northrop T-38 Talon (1961)

Top Speed: 810 MPH
Engine: (2) GE J85-GE-5;
    2,900 lbs thrust

The T-38 is a twin-engine, high-altitude, supersonic advanced jet trainer. More than 60,000 pilots have earned their wings in the T-38. In addition to teaching supersonic techniques, aerobatics, formation, night and instrument flying, and cross-country navigation, the AT-38 teaches weapons delivery as well.

Raytheon T-1 Jayhawk (1993)

Speed: 538 MPH
Engine: (2) P&W JT15D-5;
    2,900 lbs thrust

The T-1 is a version of the Beech 400A civilian aircraft and was the first new trainer aircraft purchased by the USAF in 30 years. The T-1 represented the beginning of a new era in pilot training. From the late 1950s to the early 1990s, all pilot training students received primary training in the T-37 and advanced training in the T-38. In 1993 the USAF went to specialized undergraduate pilot training with all pilots still receiving primary training in the T-37 but only fighter and bomber pilots moving on to the T-38. Pilots going to airlift and tanker units receive their advance training in the T-1 instead.

Experimental and Strategic Reconnaissance Aircraft

Bell X-1 "Glamorous Glennis" (1947)

Top Speed: 700 MPH (Mach 1.06) Achieved on 14 October 1947
Ceiling: 45,000 ft
Engine: XLR11-RM-1 rocket engine; 6,000 lbs thrust

Developed in secrecy, the XS-1 (later redesignated X-1) research aircraft was the first to explore flight beyond the speed of sound (761 MPH at sea level). Design basis for the aircraft was a .50-caliber bullet. Air Force Capt Charles E. "Chuck" Yeager became the first pilot to exceed Mach 1 when on 14 October 1947, he reached a speed of Mach 1.06 (700 MPH) at an altitude of 43,000 ft. News of the flight was not revealed until nearly two months later.

Lockheed U-2 (1956)

Speed: 430 MPH
Ceiling: >70,000 ft
Engine: P&W J75-P-13B; 17,000 lbs thrust
Range: >6,000 miles

The U-2 is a high-altitude, long-endurance reconnaissance aircraft. The U-2 operates at such high altitudes that the pilots have to wear modified spacesuits. U-2 photos alerted the United States to the deployment of Soviet missiles to Cuba at the start of the Cuban missile crisis. One U-2 (operated by the CIA) was shot down over the Soviet Union in 1960 and another was shot down over Cuba in 1962. However, improved engines and surveillance equipment make the U-2 an effective reconnaissance platform in the twenty-first century.

North American X-15 (1958)

**Top Speed: 4,520 MPH (Mach
6.72)
Ceiling: 354,200 ft (67 miles)
Engine: XLR99 rocket engine;
57,000 lbs thrust**

The X-15 was designed to explore the identifiable problems of atmospheric flight at very high speeds and altitudes. The aircraft was the first to use a throttle-able rocket engine, the first to be flown to speeds of Mach 4, 5, and 6, and the first to be flown to the lower edge of space (considered by the Air Force and National Aeronautics and Space Administration [NASA] to be an altitude of 50 miles). It demonstrated that pilots could perform at hypersonic speeds and in weightlessness. It also proved the effectiveness of reusable, piloted spacecraft, and lifting reentry upon return to atmosphere.

North American XB-70 Valkyrie (1964)

**Top Speed: 2,000 MPH
(Mach 3.08)
Ceiling: 74,000 ft
Engine: (6) YJ93-GE-3;
30,000 lbs thrust
Range: 6,000 miles**

One of the most exotic aircraft ever built. Originally conceived as a high-altitude, Mach 3-capable bomber to replace the B-52, budget cuts reduced the number of aircraft to two and the program to a research effort aimed at studying aerodynamics, propulsion, and materials used on large supersonic aircraft. On one flight above Mach 3, the recorded skin temperature on the aircraft reached 620 degrees. At Mach 3, it took the aircraft an arc of 287 miles and 13 minutes to make a 180-degree turn. The aircraft utilized the phenomenon of compression lift, where the aircraft actually rode its own shock wave. It was able to do this in part because of the wingtips that could droop in flight.

Lockheed SR-71 Blackbird (1966)

Cruise: >2,000 MPH
Ceiling: >85,000 ft
Engine: (2) P&W J58;
          32,500 lbs thrust

Like the U-2 and the F-117, the SR-71 was a product of Lockheed's famous Skunk Works. The SR-71 has the most spectacular speed and altitude performance ever recorded in an operational jet aircraft. Its powerful ramjet engines and unique shape enable the SR-71 to cruise at speeds above Mach 3. No SR-71s have ever been shot down though they have conducted high-priority reconnaissance missions in many of the world's trouble spots.

Strategic Missiles and Space Launch Vehicles

Douglas PGM-17 Thor (1959)

Length: 65 ft 10 in
Diameter: 8 ft
Weight: 109,330 lbs at launch
Engine: North American Rockedyne, liquid-propelent
      rocket; 150,000 lbs thrust
Speed: 10,250 MPH
Range: 1,725 miles
Altitude: 390 miles

The single-stage Thor intermediate range ballistic missile (IRBM) was the free world's first operational IRBM. They could deliver their nuclear payloads over 1,700 miles and were assigned to SAC until 1963. After 1963 they were used for space research and as single-stage boosters.

General Dynamics HGM-19 Atlas (1959)

Atlas I
Length: 75 ft
Width: 10 ft
Weight: 260,000 lbs
Range as ICBM: 11,500 miles

Atlas was the United States' first ICBM. It was a one and one-half stage liquid-fuel missile. It was retired from SAC in 1965 in favor of the Titan II. Hundreds of Atlas rockets have been used as space launch vehicles. Atlas I rockets were used to launch the Mercury astronauts and Atlas II rockets launched portions of the Defense Satellite Communications System in the 1990s.

Martin LGM-25 Titan (1962)

Titan I/II
Length: 98 ft 4 in/103 ft
Diameter: 10 ft (Titan II second stage as wide as first)
Weight: 221,500 lbs/330,000 lbs
Range as ICBM: 6,300 miles/9,300 miles

Titan I was a two-stage liquid-fuel ICBM. The first Titan I squadron became operational in 1962. By 1965 the Titan Is were already being replaced by Titan IIs that had twice the payload and could be fired from inside their silos. Titan IIs remained part of the US nuclear deterrent until 1987 when the remaining missiles were released for use as space launch vehicles. Modified Titan IIs launched the Gemini astronauts. Titan III was a further improvement in the Titan family used only for launching space vehicles. Titan IV, with two solid fuel boosters attached, is a heavy-lift launch vehicle able to put almost 24 tons into low Earth orbit.

Titan I

Titan IV

Titan II

Boeing LGM-30 Minuteman (1962)

Minuteman I/III
Length: 55 ft 9 in/59 ft 10 in
Diameter: 6 ft
Weight: 65,000 lbs/78,000 lbs
Range: >6,300 miles/>8,000 miles

Minuteman ICBMs are three-stage solid propellent missiles and have been the core of the US nuclear deterrent for almost 40 years. Over 500 Minuteman III missiles remain in the USAF inventory at the end of the twentieth century and there are no plans to replace them. Steady improvements since the 1960s have increased the range, accuracy, and payload of the Minuteman series and the Minuteman III can carry up to three independently targeted nuclear warheads. These missiles, however, carry only one warhead in compliance with arms limitations agreements.

Minuteman I

Minuteman III

Boeing LGM-118 Peacekeeper (1986)

Height: 71 ft
Diameter: 7 ft 8 in
Weight: 195,000 lbs
Range: >6,000 miles

The Peacekeeper was designed to take full advantage of developments in missile guidance and multiple reentry vehicle technology. Like the Minuteman, it is a three-stage, solid propellent missile but the Peacekeeper can carry up to 10 highly accurate, independently targeted nuclear warheads. In accordance with arms reduction treaties, the Peacekeeper will be phased out by 2003.

McDonnell Douglas Delta II (1989)

Delta II
Length: 125 ft 9 in
Width: 8 ft
Weight: 511,190 lbs

The first Delta rocket was launched in 1960 and used the US Air Force's Thor rocket as its first stage. Delta rockets became the workhorse of NASA's unmanned space programs. In the late 1980s the USAF started using improved Delta II rockets to launch Navstar Global Positioning System (GPS) satellites into orbit.

Boeing AGM-86 (1982)

Speed: 550 MPH
Range: >1,500 miles

The AGM-86 air-launched cruise missile represented a dramatic leap forward from previous USAF air-launched cruise missiles. It is the first such missile that can reasonably be considered a strategic weapon.

SAC used previous air-launched cruise missiles to assist in penetrating Soviet air defenses. These cruise missiles were of two basic types: decoys designed to have the same radar signature as B-52s, and the inaccurate AGM-28 Hound Dog ground attack missile. The operational concept was for the B-52s to launch their decoys just outside the range of Soviet air-defense radar and flood Soviet air defenses with more targets than they could handle. If, in spite of the decoys, the Soviet defenses found a B-52, then it would launch its nuclear-armed AGM-28s to supress enemy air defenses.

As cruise missile accuracy improved, it became possible to design these weapons to attack precision targets. The AGM-86B's sophisticated terrain-following radar enables it to fly at very low altitudes and deliver its nuclear payload with great accuracy. The AGM-86C (originally known as conventional air-launched cruise missile [CALCM]) has a conventional warhead and uses GPS guidance. AGM-86Cs were used in Operation Desert Storm and in subsequent attacks on Iraq. Because of its heavier conventional warhead, the AGM-86C has a shorter range than the B model.

Boeing AGM-129 Advanced Cruise Missile (1991)

Speed and Range are classified.

The AGM-129 is the USAF's latest generation of air-launched cruise missiles with longer range and better accuracy than the AGM-86. To enhance survivability, the AGM-129 also incorporates low-observable stealth technology.

Lockheed Corona (1960)

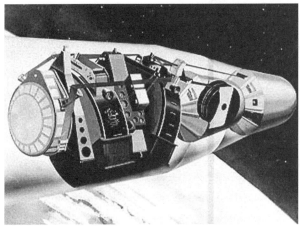

Corona Reconnaissance Satellite

C-119 recovering a Corona capsule by snagging its parachute

The Corona project was a joint Air Force/CIA program to put camera-carrying satellites into orbit over the Soviet Union and other areas of interest. The Corona satellites were launched by Thor rockets and the exposed film was ejected at a predetermined time and location in a return capsule. At about 60,000 feet, a parachute deployed and while it floated to earth, a specially modified C-119 or C-130 aircraft used a "trapeze" bar with hooks to snag the parachute in flight (photograph at right above). The aircraft then reeled the film capsule into the aircraft and returned it to base for analysis. Key to Corona's success were the enormous improvements in camera technology that enabled later Corona photographs, taken at an altitude of more than one hundred miles while traveling faster than 17,000 MPH, to identify objects as small as six feet across. More than 100 Corona missions were conducted between 1960 and 1972, when the system was retired. The systems that replaced Corona are still classified.

Lockheed Defense Meteorological Satellite Program (1960s)

Weight: 1,750 lbs
Orbital Altitude: 500 miles
Dimensions: 12 ft long, 5 ft wide
Launch Vehicle: Titan II

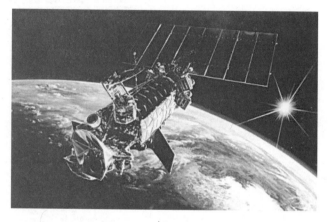

The Defense Meteorological Satellite Program (DMSP) has been collecting weather data for military operations for more than 20 years. DMSP satellites operate in polar orbit and provide continual visual and infrared imagery of cloud cover and monitor moisture, temperatures, electromagnetic fields, and other weather data. Since 1994, DMSP data has been shared with NASA and the US Commerce Department.

Defense Satellite Communications System (1971)

DSCS II/III
Weight: 1,350 lbs/2,580 lbs
Orbital Altitude: 23,230 miles
Dimensions: Height 6 ft/7 ft
Width 9 ft/6 ft
Launch Vehicle: Atlas II

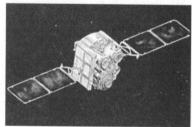

DSCS III

DSCS II

The Defense Satellite Communications System (DSCS) provides secure, high priority, command and control communications to global United States military operations. The first DSCS II was launched in 1971 and the first of the more powerful DSCS III's were first launched in 1982.

TRW Defense Support Program (1970)

Weight: 5,250 lbs
Altitude: 22,000 miles
Dimensions: 33 ft long,
22 ft wide
Launch Vehicle: Titan IV

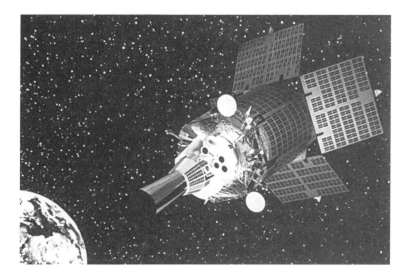

Defense Support Program (DSP) satellites provide early warning by detecting and reporting missile and space launches and nuclear detonations. Continuous improvements have kept the program up to date. These satellites provided critical early warning of Iraqi Scud launches to United States and coalition forces during Operation Desert Storm. The Space-Based Infrared System will replace the DSP in the twenty-first century.

Rockwell Navstar Global Positioning System (1978)

Weight: 3,670 lbs
Altitude: 12,000 miles
Dimensions: 11 ft long, 17 ft wide
Launch Vehicle: Delta II

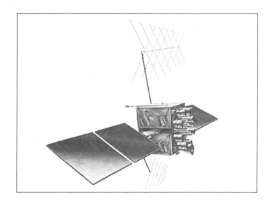

The first GPS satellite was launched in 1978, but the need for extensive testing and the large number of satellites needed to provide full coverage of the Earth (currently 24 satellites) meant that the system was not complete until 1995. Though the system was not yet fully operational, the GPS was critical to the success of United States and coalition forces in Operation Desert Storm.

Milstar Satellite Communications System (1994)

Weight: 10,000 lbs
Orbital Altitude: 25,000 miles
Dimensions: 37 ft long, 116 feet wide
    (with solar panels deployed)
Launch Vehicle: Titan IV

 Milstar is the Air Force's newest communications satellite. When fully operational, the system will consist of four satellites in geosynchronous orbits and two more in polar orbit. Each Milstar satellite serves as a smart switchboard directing traffic from terminal to terminal anywhere in the world with minimal requirement for ground controlled switching.

APPENDICES

Senior Leadership of the Early Air Forces

Air Force Chiefs of Staff

Chief Master Sergeants of the Air Force

Fighter Aces

USAF Medal of Honor Winners

Appendix A
Senior Leadership of the Early Air Forces

Chief Signal Officer
Brig Gen Adolphus V. Greely (1891–1906)
Brig Gen James Allen (1906–07)

Chief, Aeronautical Division of the Signal Corps
Capt Charles DeForest Chandler (1907–10)
Capt A. S. Cowan (1910–11)
Capt Charles DeForest Chandler (1911–12)
Lt Henry H. Arnold (1912–13)
Col Sam Reber (1913–16)

Aviation Section of the Signal Corps
Col George O. Squier (1916–17)
Lt Col John B. Bennet (1917)

Chief, Division of Military Aeronautics
Brig Gen Benjamin D. Foulois (1917)
Brig Gen Alexander L. Dade (1917–18)
Maj Gen William L. Kenly (1918)

Director of the Air Service
Maj Gen Charles Menoher (1918–21)
Maj Gen Mason M. Patrick (1921–26)

Chief of the Air Corps
Maj Gen Mason M. Patrick (1926–27)
Maj Gen James E. Fechet (1927–31)
Maj Gen Benjamin D. Foulois (1931–35)
Maj Gen Oscar Westover (1935–38)
Gen Henry H. Arnold (1938–41)

Commanding General Army Air Forces
Gen Henry H. Arnold (1941–46)
Gen Carl A. Spaatz (1946–47)

Appendix B
Air Force Chiefs of Staff

Gen Carl A. Spaatz (September 1947–April 1948)

Gen Hoyt S. Vandenberg (April 1948–June 1953)

Gen Nathan F. Twining (June 1953–June 1957)
    Served as Chairman, Joint Chiefs of Staff (August 1957–September 1960)

Gen Thomas D. White (July 1957–June 1961)

Gen Curtis E. LeMay (June 1961–January 1965)

Gen John P. McConnell (February 1965–July 1969)

Gen John D. Ryan (August 1969–July 1973)

Gen George S. Brown (August 1973–June 1974)
    Served as Chairman, Joint Chiefs of Staff (July 1974–June 1978)

Gen David C. Jones (July 1974–June 1978)
    Served as Chairman, Joint Chiefs of Staff (June 1978–June 1982)

Gen Lew Allen Jr. (July 1978–June 1982)

Gen Charles A. Gabriel (July 1982–June 1986)

Gen Larry D. Welch (July 1986–June 1990)

Gen Merrill A. McPeak (October 1990–October 1994)

Gen Ronald R. Fogleman (October 1994–September 1997)

Gen Michael E. Ryan (October 1997–present)

Appendix C
Chief Master Sergeants of the Air Force

Paul W. Airey (3 April 1967–31 July 1969)

Donald L. Harlow (1 August 1969–30 September 1971)

Richard D. Kisling (1 October 1971–30 September 1973)

Thomas N. Barnes (1 October 1973–31 July 1977)

Robert D. Gaylor (1 August 1977–31 July 1979)

James M. McCoy (1 August 1979–31 July 1981)

Arthur L. Andrews (1 August 1981–31 July 1983)

Sam E. Parish (1 August 1983–30 June 1986)

James C. Binnicker (1 July 1986–31 July 1990)

Gary R. Pfinsgton (1 August 1990–25 October 1994)

David J. Campanale (26 October 1994–1 November 1996)

Eric W. Benken (5 November 1996–30 July 1999)

Frederick J. Finch (2 August 1999–present)

Appendix D
Fighter Aces

In World War I France, Belgium, Italy, Russia, and the United States used the term *ace*. It was first used to describe the exploits of Roland Garros, who when he was captured had five kills. Thus, became the standard by which "acedom" was achieved.

World War I American Aces

Edward V. Rickenbacker	26	Frank K. Hays	6
Frank Luke Jr.	18	Donald Hudson	6
George A. Vaughn	13	Howard C. Knotts	6
Field E. Kindley	12	Robert O. Lindsay	6
Elliot White Springs	12	John K. MacArthur	6
Reed G. Landis	10	William T. Ponder	6
Jacques M. Swaab	10	David E. Putnam	6
Paul F. Baer	9	William H. Stovall	6
Thomas G. Cassady	9	Edgar G. Tobin	6
Lloyd A. Hamilton	9	Jerry C. Vasconcells	6
Chester E. Wright	9	William T. Badham	5
Henry R. Clay	8	Clayton L. Bissell	5
Hamilton Coolidge	8	Hilbert L. Blair	5
John Owen Donaldson	8	Arthur R. Brooks	5
William P. Erwin	8	Harold R. Buckley	5
Frank O. D. Hunter	8	Everett R. Cook	5
Clinton Jones	8	Charles R. D'Olive	5
James A. Meissner	8	Arthur L. Easterbrook	5
Martinus Stenseth	8	George W. Furlow	5
Wilbur W. White	8	Harold H. George	5
Howard R. Burdick	7	Charles G. Grey	5
Reed M. Chambers	7	Edward M. Haight	5
Harvey Weir Cook	7	James A. Healy	5
Jesse O. Creech	8	James Knowles Jr.	5
Lansing C. Holden	7	G. DeFreest Larner	5
Wendel A. Robertson	7	Frederick E. Luff	5
Leslie J. Rummell	7	Ralph A. O'Neill	5
Karl J. Schoen	7	John S. Owens	5
Sumner Sewall	7	Kenneth L. Porter	5
James D. Beane	6	Orville A. Ralston	5
Charles J. Biddle	6	John J. Seerley	5
Douglas Campbell	6	Victor H. Strahm	5
Edward P. Curtis	6	Robert M. Todd	5
Murray K. Guthrie	6	Remington D. Vernam	5
Leonard C. Hammond	6	Joseph F. Wehner	5

World War II US Army Air Forces Aces with at Least 14.5 Victories

Name	Victories	Name	Victories
Richard I. Bong	40	John C. Herbst	18
Thomas B. McGuire Jr.	38	Hubert Zemke	17.75
Francis S. Gabreski	28	John B. England	17.5
Robert S. Johnson	27	Duane W. Beeson	17.33
Charles H. MacDonald	27	John F. Thornell Jr.	17.25
George E. Preddy Jr.	26.83	James S. Varnell Jr.	17
John C. Meyer	24	Gerald W. Johnson	16.5
David C. Schilling	22.5	John T. Godfrey	16.33
Gerald R. Johnson	22	Clarence E. Anderson Jr.	16.25
Jay T. Robbins	22	William D. Dunham	16
Neel E. Kearby	22	Bill Harris	16
Fred J. Christensen	21.5	George S. Welch	16
Ray S. Wetmore	21.25	Donald M. Beerbower	15.5
John J. Voll	21	Samuel J. Brown	15.5
Walker M. Mahurin	20.75	Richard A. Peterson	15.5
Robert B. Westbrook	20	William T. Whisner Jr.	15.5
Thomas J. Lynch	20	Jack T. Bradley	15
Don S. Gentile	19.83	Robert W. Foy	15
Glenn E. Duncan	19.5	Ralph K. Hofer	15
Glenn T. Eagleston	18.5	Edward Cragg	15
Leonard K. Carson	18.5	Cyril F. Homer	15
Walter C. Beckham	18	John D. Landers	14.5
Herschel H. Green	18	Joe H. Powers Jr.	14.5

Capt Robert S. Johnson of the 61st Fighter Squadron shot down 27 aircraft between June of 1943 and May of 1944 in the European theater.

US Air Force Korean Conflict Aces

Joseph McConnell Jr.	16	Francis Gabreski	6.5
James Jabara	15	Donald E. Adams	6.5
Manuel J. Fernandez	14.5	George L. Jones	6.5
George Davis Jr.	14	Winton W. Marshall	6.5
Royal N. Baker	13	James H. Kasler	6
Fredrick Blesse	10	Robert J. Love	6
Harold E. Fischer	10	William T. Whisner Jr.	5.5
Vermont Garrison	10	Robert P. Baldwin	5
James Johnson	10	Richard S. Becker	5
Lonnie R. Moore	10	Stephen L. Bettinger	5
Ralph S. Parr Jr.	10	Richard D. Creighton	5
Cecil G. Foster	9	Clyde A. Curtin	5
James F. Low	9	Ralph D. Gibson	5
James P. Hagerstrom	8.5	Ivan C. Kinchloe	5
Robinson Risner	8	Robert T. Latshaw	5
George L. Ruddell	8	Robert H. Moore	5
Henry Buttlemann	7	Dolphin D. Overton III	5
Clifford D. Jolley	7	William H. Wescott	5
Leonard W. Lilley	7	Harrison R. Thyng	5

US Air Force Vietnam War Aces

Charles B. DeBellevue	6
Richard S. Ritchie	5
Jeffrey S. Feinstein	5

Appendix E
USAF Medal of Honor Winners

Name	Date of Action	Place of Action

World War I

2Lt Erwin R. Bleckley	6 Oct 1918	Binarville, France
2Lt Harold E. Goettler	6 Oct 1918	Binarville, France
2Lt Frank Luke Jr.	29 Sep 1918	Murvaux, France
Capt Edward V. Rickenbacker	25 Sep 1918	Billy, France

World War II

Lt Col Addison E. Baker	1 Aug 1943	Ploesti, Romania
Maj Richard I. Bong	Fall of 1943	Southwest Pacific
Maj Horace S. Carswell Jr.	26 Oct 1944	South China Sea
Brig Gen Frederick W. Castle	24 Dec 1944	Liege, Belgium
Maj Ralph Cheli	18 Aug 1943	Wewak, New Guinea
Col Demas T. Craw	8 Nov 1942	Port Lyautey, French Morocco
Lt Col James H. Doolittle	18 Apr 1942	Tokyo, Japan
SSgt Henry E. Erwin	12 Apr 1945	Koriyama, Japan
2Lt Robert E. Femoyer	2 Nov 1944	Merseburg, Germany
1Lt Donald J. Gott	9 Nov 1944	Saarbrucken, Germany
Maj Pierpont M. Hamilton	8 Nov 1942	Port Lyautey, French Morocco
Lt Col James H. Howard	11 Jan 1944	Oschersleben, Germany
2Lt Lloyd H. Hughes	1 Aug 1943	Ploesti, Romania
Maj John L. Jerstad	1 Aug 1943	Ploesti, Romania
Col Leon W. Johnson	1 Aug 1943	Ploesti, Romania
Col John R. Kane	1 Aug 1943	Ploesti, Romania
Col Neel E. Kearby	11 Oct 1943	Wewak, New Guinea
2Lt David R. Kingsley	23 Jun 1944	Ploesti, Romania
1Lt Raymond L. Knight	25 Apr 1945	Po Valley, Italy
1Lt William R. Lawley Jr.	20 Feb 1944	Leipzig, Germany
Capt Darrell R. Lindsey	9 Aug 1944	Pontoise, France
SSgt Archibald Mathies	20 Feb 1944	Leipzig, Germany
1Lt Jack W. Mathis	18 Mar 1943	Vegesack, Germany

USAF Medal of Honor Winners (Cont.)

Name	Date of Action	Place of Action
Maj Thomas B. McGuire Jr.	25-26 Dec 1944	Luzon, Philippines
2Lt William E. Metzger Jr.	9 Nov 1944	Saarbrucken, Germany
1Lt Edward S. Michael	11 Apr 1944	Brunswick, Germany
2Lt John C. Morgan	28 Jul 1943	Kiel, Germany
Capt Harl Pease Jr.	7 Aug 1942	Rabaul, New Britain
1Lt Donald D. Pucket	9 Jul 1944	Ploesti, Romania
2Lt Joseph R. Sarnoski	16 Jun 1943	Buka, Solomon Islands
Maj William A. Shomo	11 Jan 1945	Luzon, Philippines
Sgt Maynard H. Smith	1 May 1943	St. Nazaire, France
2Lt Walter E. Truemper	20 Feb 1944	Leipzig, Germany
Lt Col Leon R. Vance Jr.	5 Jun 1944	Wimereaux, France
TSgt Forrest L. Vosler	20 Dec 1943	Bremen, Germany
Brig Gen Kenneth N. Walker	5 Jan 1943	Rabaul, New Britain
Maj Raymond H. Wilkins	2 Nov 1943	Rabaul, New Britain
Maj Jay Zeamer Jr.	16 Jun 1943	Buka, Solomon Islands

Korea

Name	Date of Action	Place of Action
Maj George A. Davis Jr.	10 Feb 1952	Sinuiji-Yalu River, N. Korea
Maj Charles J. Loring Jr.	22 Nov 1952	Sniper Ridge, N. Korea
Maj Louis J. Sebille	5 Aug 1950	Hamch'ang, S. Korea
Capt John S. Walmsley Jr.	14 Sep 1951	Yangdok, N. Korea

Vietnam

Name	Date of Action	Place of Action
Capt Steven L. Bennett	29 Jun 1972	Quang Tri, S. Vietnam
Col George E. Day	While POW	
Maj Merlyn H. Dethlefsen	10 Mar 1967	Thai Nguyen, N. Vietnam
Maj Bernard F. Fisher	10 Mar 1966	A Shau Valley, S. Vietnam
1Lt James P. Fleming	26 Nov 1968	Duc Co, S. Vietnam
Lt Col Joe M. Jackson	12 May 1968	Kham Duc, S. Vietnam
Col William A. Jones III	1 Sep 1968	Dong hoi, N. Vietnam
A1C John L. Levitow	24 Feb 1969	Long Binh, S. Vietnam
A1C William H. Pitzenbarger	11 Apr 1966	S. Vietnam
Capt Lance P. Sijan	While POW	
Lt Col Leo K. Thorsness	19 Apr 1967	N. Vietnam
Capt Hilliard A. Wilbanks	24 Feb 1967	Dalat, S. Vietnam
Capt Gerald O. Young	9 Nov 1967	Da Nang area, S. Vietnam

The Medal of Honor is the highest military award for bravery that can be given to any individual in the United States of America. Established by joint resolution of Congress, 12 July 1862 (amended by Act of 9 July 1918 and Act of 25 July 1963). The first Medals of Honor awarded went to six Union Army volunteers, members of Andrews's (or Mitchel's) Raiders. These men attempted to disrupt the Confederate rail lines between Atlanta and Chattanooga. Disguised as civilians, the 19 soldiers, along with civilians James Andrews and William Campbell, captured the locomotive *General* at Big Shanty, Georgia. Chased by the persistent conductor of the *General* the raiders charged north, attempting to burn bridges and destroy track along the way. Less than 20 miles south of Chattanooga the "Great Locomotive Chase" ended. All of the raiders were captured and eight were tried and executed. On 25 March 1863, six of the raiders arrived in Washington after parole from a Confederate prison, and these six were the first to be presented Medals of Honor by Secretary of War Stanton.

GPO U.S. GOVERNMENT PRINTING OFFICE: 2005–735–265